TRADE UNION BEHAVIOUR, PAY-BARGAINING, AND ECONOMIC PERFORMANCE

The Trade Union Institute for Economic Research, FIEF, is a foundation established in 1985 by Landsorganisationen, the Swedish trade union confederation. FIEF's objective, as defined in its constitution, is to 'deepen the academic economic debate through the promotion of enduring research'.

FIEF Studies in Labour Markets and Economic Policy will be published once a year. The series will provide a forum for outstanding scholars to publish applied, policy-oriented research with generous space available. The length of the papers should be between 40 and 60 pages which allow background surveys of theory, and a review of empirical research. The papers should also contain original contributions either through extensions and/or of empirical analysis.

Normally, two conferences are organized around the papers to be published in the *FIEF Studies*. After the first conference, papers are revised and a final conference is held with FIEF's panel of advisors and specially invited researchers in the field covered by the paper.

Trade Union Behaviour, Pay-Bargaining, And Economic Performance

R. J. FLANAGAN

K. O. MOENE

M. WALLERSTEIN

CLARENDON PRESS · OXFORD

1993

Oxford University Press, Walton Street, Oxford OX2 6DP
Oxford New York Toronto
Delhi Bombay Calcutta Madras Karachi
Kuala Lumpur Singapore Hong Kong Tokyo
Nairobi Dar es Salaam Cape Town
Melbourne Auckland Madrid
and associated companies in
Berlin Ibadan

Oxford is a trade mark of Oxford University Press

Published in the United States
by Oxford University Press Inc., New York

British Library Cataloguing in Publication Data
Data available

Library of Congress Cataloging-in-Publication Data
Trade union behaviour, pay bargaining, and economic performance / R.J. Flanagan . . . [et al.].
p. cm. — (FIEF studies in labour markets and economic policy)
Includes bibliographical references and index.
1. Trade-unions—Congresses. 2. Collective bargaining—Congresses. 3. Wages—Congresses. 4. Wages and labor productivity—Congresses. I. Flanagan, Robert J. II. Series.
HD6483.T65 1993 331.8′—dc20 92-41369

ISBN 0–19–828798–4

1 3 5 7 9 10 8 6 4 2

Typeset by Best-set Typesetter Ltd., Hong Kong
Printed in Great Britain
on acid-free paper by
Bookcraft (Bath) Ltd.
Midsomer Norton, Avon

Contents

Participants in the conference on papers in this volume

Susanne Ackum-Agell	Ph.D. Student, Uppsala University
Villy Bergström	Director of the Trade Union Institute for Economic Research, FIEF
Lars Calmfors	Professor, Institute for International Economic Studies, IIES, University of Stockholm
Per-Anders Edin	Ph.D., Uppsala University
Robert J. Flanagan	Professor, Standford University
Siv Gustafsson	Professor, University of Amsterdam
Magnus Henrekson	Econ. Dr., Trade Union Institute for Economic Research, FIEF
Douglas A. Hibbs, Jr.	Professor, Trade Union Institute for Economic Research, FIEF
Michael Hoel	Professor, Oslo University
Bertil Holmlund	Professor, Uppsala University
Assar Lindbeck	Professor, Institute for International Economic Studies, IIES, University of Stockholm
Per Lundborg	Docent, Industrial Institute for Economic and Social Research, IUI
Christian Nilsson	Ph.D., Uppsala University
Henry Ohlsson	Ph.D., Uppsala University
Åsa Rosén	Ph.D., London School of Economics
Nils Henrik Schager	Chief Economist, The Swedish National Agency for Government Employers, SAV
Claes Henrik Siven	Professor, University of Stockholm
Per Skedinger	Ph.D., Trade Union Institute for Economic Research, FIEF

Hans T. Söderström	Executive Director, Center for Business and Policy Studies, SNS
Eva Uddén-Jondal	Licenciate in Economics, Stockholm School of Economics
Alistair Ulph	Professor, University of Southampton
Magnus Wikström	Ph.D. Student, University of Umeå
Johnny Zetterberg	Ph.D., Uppsala University
Thomas Östros	Ph.D. Student, Uppsala University

List of Figures

List of Tables

Introduction

This volume of *FIEF Studies in Labour Markets and Economic Policy* consists of two purely theoretical papers. The paper by Professor Robert Flanagan deals with the process of policy formulation within unions and the one by Karl Ove Moene, Michael Wallerstein, and Michael Hoel with the bargaining structure in the labour market.

Professor Flanagan was one of the contributors to the study of the Swedish economy by the Brookings Institution in 1986. Visiting Sweden, Flanagan was struck by the emphasis on equalization of wages, especially by Sweden's blue-collar unions. This tendency is observed, albeit less clearly, within other Swedish unions as well as in unions in other countries.

The models used to analyse wage-formation mostly disregard unions' policy to promote the so-called solidaristic wage policy. The typical arguments in a union utility function are real wages and employment and nothing more. This puzzled Flanagan and aroused his interest in the internal decision-processes of trade unions.

Union membership is heterogeneous, made up of people of different ages, occupations, and locations. How are the different interests stemming from these heterogeneities aggregated to a union policy? There are also principal–agent problems regarding membership and leadership of unions, seldom treated in the labour-market literature.

In his paper Flanagan uses collective-choice analysis to shed light on the decision-making within trade unions. He tries classical voting models with individual and representative voting and he invokes the median voter theory to explain the formation of trade union utility functions.

Flanagan's analysis leads to paradoxical results. Multidimensional voting models imply unstable union policies, with shifting emphasis over time on employment, wages, insurance issues, and so forth. Such instability, he argues, is seldom observed empirically. Some may object to this though, and say that shifting objectives have occurred recently. One example is the wage concessions made by US unions in crisis industries, such as the car industry. Another example is the decreasing emphasis on solidaristic wage policy after the early 1980s by Sweden's blue-collar unions.

The median voter model on the other hand does produce equilibrium outcomes based on an ordering of preferences by seniority. This may fit US labour markets but not European, where work-sharing rather than wage concessions have been used to increase job security of members.

In the second part of his paper Flanagan analyses the institutional structure of unions. Are there internal political processes and norms that secure stable outcomes of the political process and the voting mechanism? The internal political process of unions may be arranged to produce greater uniformity of preferences of voting members or reductions in the number of issues that are submitted to voting among union members.

One means to limit the number of dimensions subject to the bargaining process is centralization of bargaining (although this would increase heterogeneity of constituent preferences). This structural question is also the subject of the second paper in this volume of *FIEF Studies*.

Large international variation in macro-economic performance across countries after the first oil shock in the early 1970s generated considerable interest in the structure of bargaining in different countries. In the Nordic countries, which since the 1950s have had a strong central component in wage-bargaining, this structure came under attack from the employers' organizations. The debate on this issue is vivid, especially in Sweden.

There have been many empirical investigations of the consequences of bargaining structures, such as international differences in the trade-off between employment and inflation and the frequency of strikes. Empirical analyses in this field suffer from lack of data (few observations), multi-dimensional variables—it is seldom easy to classify a certain country as centralized or decentralized—and a large number of potentially influential factors behind an observed phenomenon as broad as 'economic performance'. Results of empirical research are often vague and uncertain. Because of these difficulties Moene, Wallerstein, and

Hoel concentrate on what economic theory has to say about the effect of different bargaining structures on economic performance.

The paper by Moene, Wallerstein, and Hoel starts out in the framework of models where unions can impose wages, subject to a demand for labour constraint. These models describe union wage demands rather than bargaining *per se*, which is studied in Chapter 12. The model framework starts in a simple world, and then is made increasingly more complex by bringing in externalities, so that wages influence prices and other variables. In these different environments the authors investigate the consequences of different levels of union centralization. The impact of the level of centralization on the trade-off between wages and employment is the basis for the discussion in the paper.

The Nordic system of negotiations typically has been centralized to national unions but allowed negotiations on the industry level and the firm or production unit level as well. This gives rise to so-called wage drift, which used to make up some 50 per cent of the total of wage increases in Sweden. This negotiation structure is analysed within a bargaining framework. At the end, the authors discuss the bargaining system itself as an object of negotiation.

The bargaining literature consists of a multitude of model approaches with different assumptions, and emphasizes different aspects of labour-market problems. One strength of the paper by Moene, Wallerstein, and Hoel is the simplicity of their model approach and its coherence. The authors succeed in applying basically the same model approach to the great host of problems they set for themselves.

FIEF organized a conference on the papers in this volume. Invited speakers included Alistair Ulph, and his discussion of Flanagan's paper appears in the book, as does Lars Calmfors's discussion of the Moene, Wallerstein, and Hoel paper. Assar Lindbeck was an invited speaker on both papers, and his comments are also included.

Readers may not find definitive answers to the questions of how unions formulate their policy stance and what union policy will be, nor to which bargaining structure is superior in terms of economic efficiency. But from this volume of *FIEF Studies* you will get a better understanding of why there are no clear-cut, definitive answers at this stage in the development of labour economics.

Villy Bergström
Director, FIEF

PART I

Can Political Models Predict Union Behaviour?

1
Introduction

In their decision-making, trade unions resemble governments more than firms. Union leaders are elected directly or indirectly by the union membership, bargaining goals are established by participatory mechanisms, tentative collective-bargaining agreements must be ratified by a vote of the membership or a body elected by the membership, and strikes must be authorized by membership vote. Despite these central institutional features, most theories of trade union behaviour take little or no account of the collective choice process that drives decision-making in unions. The emergence of a union's objectives from the internal voting procedures through which conflicts among union members and between union members and their leaders are resolved is a *deus ex machina* in most modern studies of trade unions.

When empirical comparisons between union and non-union employment relationships can be made, they show distinctive differences between employment arrangements established in a hierarchical (non-union) setting and those that emerge from a collective-bargaining setting in which voting processes establish employees' objectives. While hierarchical determination of employment arrangements may dominate in some labour markets, such as those in the United States and Japan, collective choice would appear to dominate in most Western European countries, where the combined effects of collective bargaining and legal extension of collective-bargaining outcomes to non-members virtually eliminate a truly non-union sector. Failure to understand

Robert J. Flanagan is Professor of Labor Economics, Graduate School of Business, Stanford University. The author is grateful for the financial support of the Graduate School of Business at Stanford and for the hospitality of the Research School for the Social Sciences, Australian National University and the Netherlands Institute for Advanced Study (NIAS) where portions of this essay were written. Helpful comments were received from David Baron, Jonathan Bendor, Assar Lindbeck, John Pencavel, Alistair Ulph, and participants at the 1991 FIEF conference in Stockholm and at seminars at NIAS, the Catholic University in Louvain, and the University of Leiden.

the determination of employment arrangements in a collective choice framework would seem to inhibit understanding of employment arrangements in large segments of the labour market.

The purpose of this essay is to explore the possible gains from applying modern collective choice analyses to the formation of trade union objectives. The essay examines whether and under what circumstances it is possible for union members to form a collective goal and the ability of union leaders to alter that goal. Much of the essay applies models developed recently by political scientists for formal study of (American) legislative institutions to a union setting. The essential question addressed by these models is whether particular voting arrangements produce equilibria that enable secure predictions about the objectives of collective choice institutions. Unlike many recent studies of unions, this essay does not focus on the outcome(s) of a bargaining process, but rather on the mechanisms through which the preferences that surely influence bargaining outcomes are formulated in the first place. Thus, the level of analysis is the organization and its internal voting procedures rather than the market and bargaining procedures.

Chapter 2 briefly reviews the nature and limitations of modern analyses of trade union behaviour when viewed from a collective choice perspective. Chapter 3 develops the implications of classical voting models for the determination of union objectives and notes the paradox that the chaotic behaviour predicted by these models is not generally encountered in reality. The remaining chapters explore methods of resolving this paradox. Chapter 4 reviews applications of the well-known median voter model to certain micro and macro aspects of union behaviour. Chapter 5 examines the effects of institutional constraints on the exercise of individual preferences in voting. The key issue here is the extent to which chaos is transformed into equilibrium when specific institutions of union decision-making are overlaid on membership preferences. The effects of jurisdiction, bargaining structure, and voting rules are considered. In Chapter 6, the essay considers the extent to which union leaders can, by virtue of their position and authority to negotiate tentative agreements, achieve results different from the preferences of the median member. This raises issues of agenda control and principal–agent relationships. The essay concludes with a stock-taking of the possibilities and problems inherent in the application of political models to unions.

2

Trade Unions: Models and Institutions

Nowhere in economics have the objectives of an agent been modelled with less attention to foundations than in the study of unions. In a setting in which collective choices are the rule of the day, most models evade the question of how trade union policy evolves from the preferences of members and leaders through internal union political processes. Notwithstanding the value of parsimony in economic analysis, it is natural to question the output of models in which the objectives of one of the key actors are treated so casually. This section contrasts the treatment of union objectives in the more popular bargaining models with key institutional features of decision-making in unions. This contrast provides the point of departure for the examination of explicitly political models of union behaviour in the rest of the essay.

Two formulations of union objectives have emerged in economic models of union bargaining. Each fails to capture the collective choice aspect of union decision-making in significant ways. Under right-to-manage formulations of union–employer interactions, the union's objectives are summarized by a utility function in which a typical member's welfare increases in the real wage rate and employment (and possibly other variables), while the profit-maximizing employer's interests are summarized by the labour demand curve. In the monopoly union version of the right-to-manage formulation, the union chooses the wage rate that maximizes its utility with full knowledge of the employer's demand curve. The employer then chooses the profit-maximizing employment level corresponding to that wage from the labour demand curve. (The fact that many labour agreements specify wage levels but leave employment decisions to management is consistent with the model, although employer disinterest in the wage is not.) Under less-extreme versions, the employer chooses the employment level after both parties negotiate the wage. Among other things, the model 'predicts' that unions should give weight to employment as well as the wage, and studies of a limited number of unions support this. These models, particularly the monopoly union version, are easy to criticize for their lack of correspondence to obvious details of the collective bargaining process, and many have done so.[1]

In the present essay, however, the focus is on the union objective function used in the model, which is variously described as representing the welfare of 'a typical union member' or the objectives of the union leader. Each of these characterizations presents problems in a foundational sense. Explaining union behaviour on the basis of the utility function of the 'typical' union member is admissible when all union members have identical preferences. But it is a rare union whose policy is chosen by unanimous consent. Instead, union membership is marked by its heterogeneity and can encompass groups with significant conflicts of interest. How are the heterogeneity mediated and the conflicts resolved to produce something that might be called a 'typical' membership objective? With rare exceptions (noted later) economic models of collective bargaining have been silent on these issues. Moreover, except in a world of dictatorial unions, the 'union leader's objectives' interpretation is also uneasy, for it resolves none of the principal–agent issues that it raises. To what extent can a leader's objectives depart from those of the members that elect him? Once a leader must consider the membership's goals when formulating his own objectives, all the issues raised above resurface.

As an alternative, other bargaining models have adopted an individualistic utilitarian form of union objectives, in which the union is concerned with the sum of members' utility levels. In implementation, two categories of members are recognized—employed and unemployed—with the former receiving the utility associated with the union wage, while the latter receive the utility associated with their best alternative (e.g. unemployment benefits, leisure, or a non-union wage). This approach admits limited heterogeneity—employed and unemployed members have different utility levels—but differences that typically exist within each of these groups are ignored. This specification also

[1] See critical reviews by Farber (1986), Holmlund (1989), Oswald (1985), and Pencavel (1985; 1991), for example, which note the absence of a bargaining process, the employer's passive role in negotiations, and the absence of considerations of bargaining power.

implicitly assumes that unemployed members vote on union policy.

One might add to these criticisms that the utility functions used in bargaining models are rather sparse, ignoring, for example, the near universal goal among unions of achieving greater equality. In application, moreover, the models yield few predictions about major changes in union policy, such as wage concessions, shifts to riskier compensation arrangements such as profit-sharing or employee ownership, and union mergers. Ultimately one would like some solid foundations for the emergence of these objectives.

In contrast, two distinctive political processes influence the economic activities of most trade unions. (Most of the details that follow reflect arrangements in US unions.) One process centres on the union's periodic national conventions, which elect the union's national officers, establish their compensation, and set the organization's administrative policies and procedures. The second process establishes and implements, subject to collective-bargaining outcomes, the union's economic policies. The two political processes rarely overlap, because bargaining and membership representation activities are generally (although not invariably) more decentralized than administrative activities.

With regard to the first process, the union convention is the highest tribunal of most unions and comes closest to traditional legislative institutions. Conventions address changes in a national trade union's constitution, finances, organizational issues, and the election of officers. Under current US law, delegates to the conventions must be chosen by open democratic procedures. There are several standing committees, which screen and present proposals to the floor for a vote, and there are rules governing the procedures by which committee proposals may be amended or voted upon by the convention floor. In a distinct minority of unions, convention proposals may be put to a general referendum of the membership.

A different set of voting procedures governs the formulation and implementation of bargaining goals. Negotiations are typically conducted by a bargaining committee, led by elected union officials. Based on an assessment of membership preferences, the committees formulate bargaining goals, negotiate with a management bargaining team, and in the vast majority of negotiations reach a tentative agreement on a new collective bargaining agreement. Before taking effect, the tentative agreement must be ratified. At this stage there is some variation in the practice of individual unions. Some require a ratification vote of the entire membership, while others permit ratification by an intermediate body elected by the membership. Regardless of the exact procedure, ratification requires a majority vote. Strikes do not automatically follow a failure to ratify a tentative agreement for two reasons. Labour and management may continue efforts to negotiate an acceptable agreement, if they wish. Moreover, strike actions require a separate vote of the membership and generally require a larger majority than ratification votes. Unions may therefore reject proposed agreements without committing themselves to the costs of a strike.

The remainder of the essay considers the outcomes of mediating union membership heterogeneity and differences in the objectives of union members and their leaders through these collective choice processes. Despite the more appealing foundations, it will become apparent that progress on union objectives remains difficult in a collective-choice framework.

3

Classical Voting Models and Union Behaviour

The previous chapter stressed the importance of membership heterogeneity and collective choice processes in the formulation of union objectives. This chapter applies classical voting models of unrestricted majority rule to two observed procedures for formulating union policy or ratifying contracts—vote of the membership and vote of a representative bargaining committee.

Individual Voting

Unions fit the simple structure of classical voting models as easily as legislative institutions. In the most general model several voters choose among several issues. The analogy is union members voting over the elements of a proposed collective-bargaining agreement. Each union member is assumed to be motivated by self-interest (rather than, say, group solidarity) and to vote sincerely. (Possibilities for strategic voting are discussed later.) A majority vote of the union membership ratifies a proposed contract.

The example in Fig. 3.1 demonstrates that the interests of a union do not arise in a predictable way from the preferences of its members. Here three union members are voting on contract proposals. (The limitations on membership and scope of contract are for expository simplicity only.) Member 1 prefers a pension over a wage increase over job security guarantees. Members 2 and 3 rank the issues differently. Individual preferences are assumed to be consistent (transitive), so that Member 1 prefers a pension over job security guarantees and so on for the other members. What happens when the three members vote to determine union policy? In pairwise voting, a majority (members 1 and 2) will choose a pension over a wage increase, another majority (members 1 and 3) will prefer a wage increase to better job security, but a third majority (members 2 and 3) will also

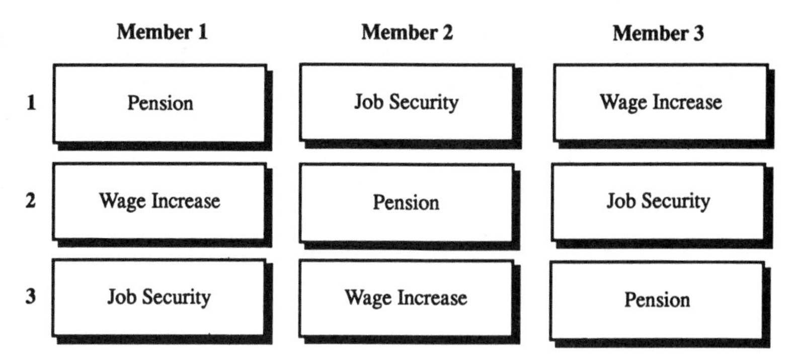

FIG. 3.1 Preference Rankings of Union Members

prefer job security over a pension! Paradoxically, the fact that each union member can rank the desirability of different elements of a contract provides no guarantee that there is an unambiguous best choice for the union. Each of the proposals is vulnerable to defeat in an unrestricted majority rule contest.[1]

Representative Voting

The situation does not improve when contract ratification occurs by vote of elected representatives of the union membership (e.g. an elected bargaining committee) rather than the members themselves. Consider the case of a three-member elected ratification committee considering a contract consisting of wage and non-wage provisions (illustrated in Fig. 3.2). Each point in the figure is a potential union contract (wage and non-wage cost level). Which contract will be ratified by elected representatives of the membership? Elected representatives must consider the diverse preferences of their constituencies if they are to retain their positions. The ideal point for each representative captures the preference of her constituency (e.g. different union locals or age groups) on the wage and non-wage dimensions of the contract. In

[1] This is just an application of Condorcet's voting paradox. Arrow's (1951) famous impossibility theorem moved beyond pairwise majority voting to establish the broader result that there are no guarantees that any democratic social institution will aggregate preferences of individual members in a way that produces an unambiguous group objective.

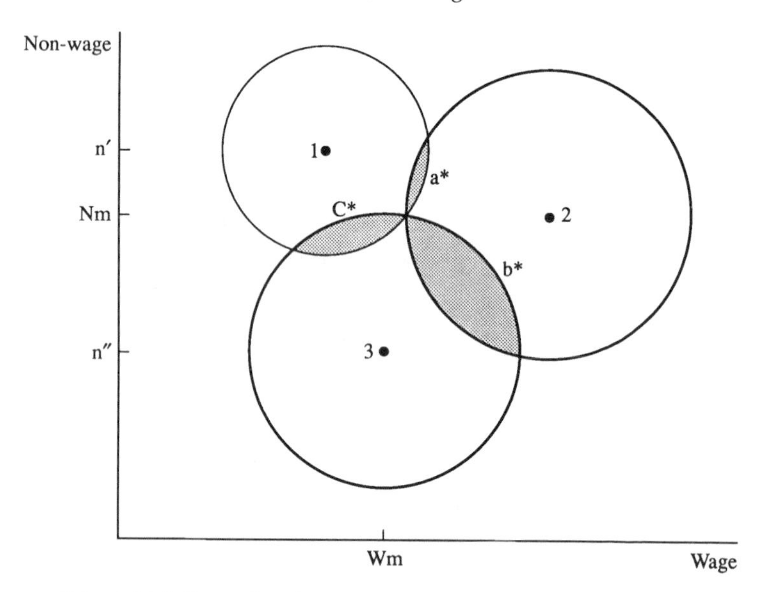

FIG. 3.2 Multi-dimensional Membership Preferences

Figure 3.2, the ideal points for three elected representatives of the membership are represented by 1, 2, and 3. Deviations from an ideal point in any direction reduces the utility (erodes the electoral support) of the representative. Each representative's satisfaction with a proposed contract package depends on the distance of the proposal from her ideal point. The relevant indifference curves for each member are therefore circles around the member's ideal point, with proposals on curves that are closer to the ideal point preferred over those on more distant curves.[2]

At first glance, contract C* may seem like the equilibrium solution because it represents the median of the preferences on each dimension. This is not a stable equilibrium, however. In comparison to C*, for example, members 1 and 2 will prefer contract a* (which puts both on more favourable indifference

[2] These are Euclidean preferences, commonly used in spatial models of legislative choice (Krehbiel 1989). The use of circular indifference curves imposes the assumption of separable preferences—that a union member's preferences on one issue (e.g. wages) do not depend on what happens on other issues (e.g. pensions). Non-separable preferences can be represented in the same framework by ellipse-shaped indifference curves rotated around ideal points.

curves), and members 2 and 3 will prefer b* for the same reason. More generally, under unrestricted, representative, majority rule, there is again no contract proposal that is invulnerable to defeat by some other proposal! (McKelvey 1976). Consider contract proposal C* again. The cross-hatched areas in Fig. 3.2 describe the alternative contract proposals that would receive majority support over C*. (This is the 'win set' of C*—the set of proposals that would defeat C* in a majority rule vote.) Proposals in these areas place two of the three representatives on more preferred indifference contours than proposal C*. (Representatives 1 and 2 are better off with any proposal in the upper win set, while members 2 and 3 are better off with any proposal in the lower win set.) But there is also a win set that dominates any proposal within these areas, and more generally, any proposal. (To check this, consider the stability of the contracts a* and b*. The same analysis and arguments apply.) With sufficiently long voting agendas, a union could 'wander' via a sequence of votes from any starting-point, such as the existing contract, to any alternative contract.

Implications

These results have fundamental implications for unions and other democratic institutions. Even with knowledge of the preferences of individual union members, there is no clear prediction about union policy, because no proposed bargaining demand can claim a majority against all other proposals. Any possible outcome will be dominated by some other outcome; each union member or elected representative is potentially the pivotal voter. Taken at face value, the notion of a 'typical' union member, employed or unemployed, is absurd in this most basic of collective choice settings.

Since there is no guarantee that any proposal stands highest on a union's preference ordering, voting 'cycles' are a real possibility over time as first one and then another coalition of members forms to determine policy. One interpretation of classical voting models is therefore that union policy will be chaotic and will exhibit little consistency over time. Unions would either have difficulty determining policy, or policy would vary substantially from one negotiation to another as different membership coalitions gained and lost ascendancy.

Those looking for special insights into union behaviour must be discouraged by the profound indeterminacy of the bare-bones classical voting models of group activity. Rather than producing sharper predictions, it indicates that 'anything can happen'. Yet, the instability of outcomes predicted by unrestricted majority rule is not generally observed in unions (or legislatures). Both bargaining proposals and the content of the collective bargaining agreements are fairly stable over time. Although significant shifts in policy direction occasionally occur in some unions—witness the development and spread of wage concession agreements during the 1980s in the United States—we do not observe wild fluctuations in either the basic objectives of unions or their implementations in collective bargaining. Indeed, the inertia in the policy objectives of most unions suggests a bias toward the status quo. Thus a viable political model of unions must explain not only the stability of union policy but also the relative invulnerability of the status quo.

The discussion in this chapter has now come full circle. Analytically, the notion of a typical union member, representative of the organization's objectives, seems dubious in this setting, but the empirical importance of the status quo suggests that some notion of a typical member may have merit. Unfortunately, this still leaves the question of which member is typical and why unanswered. These questions are pursued in subsequent chapters by exploring the nature of voting equilibria that do exist in more restricted settings.

4

The Descriptive Power of the Median Voter Model

In contrast to the chaotic world of unrestricted majority rule, the familiar median voter model generates stable equilibria that have been used to rationalize some micro and macro aspects of union behaviour. The median voter model assumes that union members have single-peaked preferences (i.e. each member's utility declines with outcomes that are farther in either direction from his preferred outcome) and vote on only one element of a contract. Under these circumstances, Black (1958) showed that the contract proposal that corresponds to the median of the distribution of membership preferences will receive majority support.[1] The median union member is the pivotal voter and his preferences determine union policy. An implication of the median voter model is that changes in union policy reflect shifts in the position of the median member (via changes in membership or changes in the views of the present median voter). The status quo is therefore vulnerable to changes in the preferences or identity of the median voter.

At the conceptual level, the median voter model rests on requirements that seem rather special. In particular, voting occurs on one issue at a time, and once this assumption is abandoned, we return to the setting illustrated earlier by Fig. 3.2.

At the empirical level, however, median voter arguments can rationalize several micro and macro aspects of union behaviour. Differences in employment arrangements between union and non-union firms in the United States broadly correspond to the notion that the preferences of the median worker determine employment objectives in the former sector, while the preferences of the marginal worker determine arrangements in the

[1] Furthermore, if candidates for union office choose platforms competitively to maximize the votes they receive, the platforms of candidates for union office should converge to the preferences of the median union member. All candidates should offer essentially the same platform.

latter sector. (The legal extension of collective bargaining agreements to non-union situations and the breadth of bargaining structures generally preclude such comparisons in Europe.) Freeman and Medoff (1984) provide many examples based on the idea that the median voter-worker is likely to be older than the marginal (most mobile) worker. This may explain the greater frequency and generosity of pension plans in the union sector, for example. Older workers will be partial towards the use of seniority (rather than random selection or, in the case of lay-offs, work-sharing) to curb employer discretion in allocating internal economic opportunities, and seniority arrangements are most common in unionized employment. Collective preference for greater pay equality in a world of log-normal wage distributions also fits the model, although the question of whether the wage distributions relevant for most bargaining units are in fact log normal is unproved, and there may be other (e.g. insurance) motives for preferring less-dispersed wages (Agell & Lommerud 1990).

Of these, the collective preference for seniority has the most far-reaching implications. It explains greater preference for fixed wage payments over pay that is contingent on firm performance in the union sector, for example. When lay-offs occur according to reverse seniority, and the normal range of demand shocks threatens a minority of worker-members, the more senior majority recognizes a fixed wage for the fixed income that it is and rejects the riskier income stream implied by profit-sharing and related compensation schemes. Fixed-wage systems may also be preferred because they provide a measure of wage equality among employed union workers, although they raise the risk faced by the least senior workers. Economists have also invoked lay-offs by reverse seniority to demonstrate that efficient bargains can in principle lie on the employer's demand curve (Oswald 1985; 1987).

At the macro level, the median voter model apparently explains the seemingly disparate patterns of union wage behaviour in the United States and Europe in the face of increasing unemployment during the 1980s. In the United States, a number of developments during the 1970s produced downward pressure on union employment. The average union relative wage increased from about 112 to 125 per cent of the average non-union wage between the late 1960s and early 1980s. During this period Government deregulation of several industries removed important barriers to non-union entry, and in the early 1980s the country experienced the deepest recession since the Great Depression. Faced with these developments, US unions initially maintained their standard agreements, which typically provided for a sequence of non-contingent wage increases over the two- or three-year life of the contract. Substitutions in production and consumption produced a huge quantity adjustment, and union membership fell by some 4 million workers between 1974 and 1984. (By the late 1980s, private-sector union density had dropped to 12 per cent.) The fact that the initial burden of adjustment fell on employment rather than wages is consistent with some 'non-political' bargaining models (e.g. McDonald & Solow 1981).

In the face of the huge quantity adjustments in unionized labour markets, US unions eventually accepted wage moderation, however. When it finally arrived, the deceleration of wages was precipitous. (See Figs. 4.1 and 4.2, which trace the development of union money wage-changes in major collective bargaining units (covering 1,000 or more workers) in manufacturing and

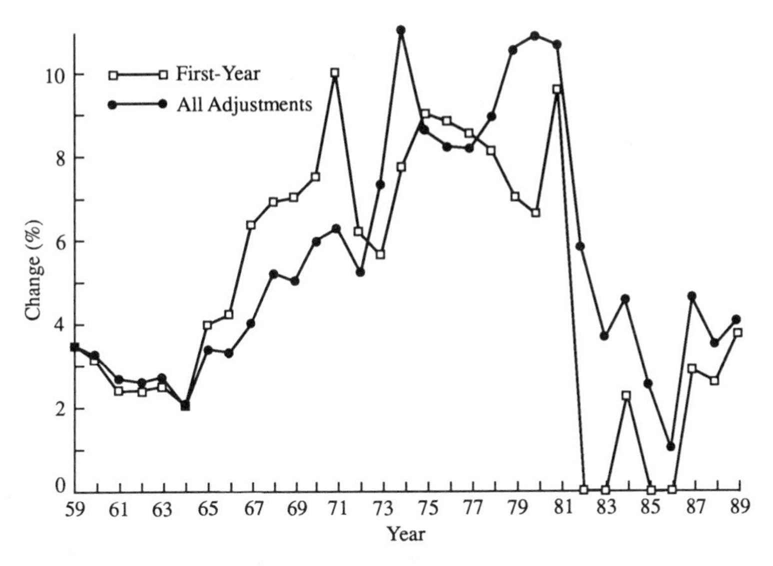

FIG. 4.1 Union Wage Changes, Manufacturing

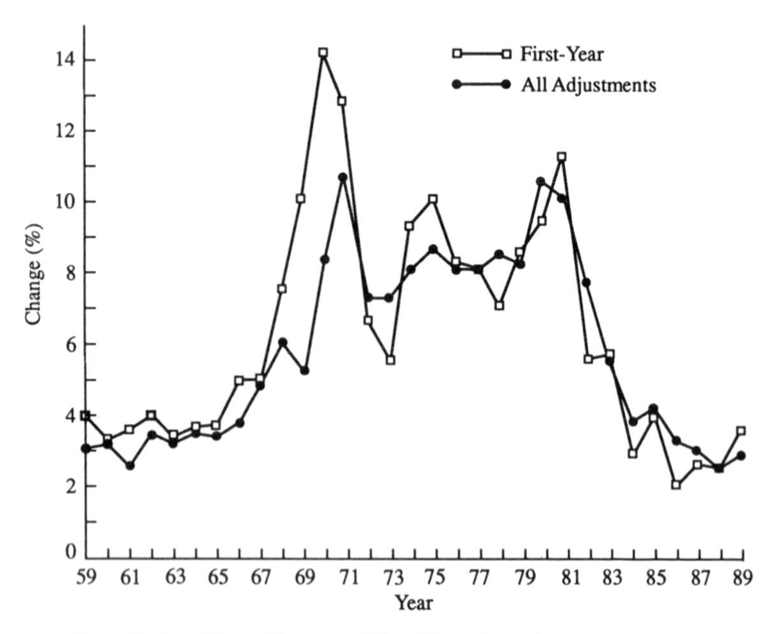

FIG. 4.2 Union Wage Changes, Non-Manufacturing

non-manufacturing respectively. These graphs compare median wage changes scheduled for the first year of recently negotiated agreements with median union wage 'adjustments', which also include the effects of cost-of-living adjustments and deferred wage increases provided for by contracts negotiated in earlier years.) First-year union wage increases in manufacturing collapsed in 1981–2, and for four of the five years beginning in 1982, major collective-bargaining agreements with wage decreases or no wage change outnumbered agreements with wage increases. As a result, union wage levels declined rapidly relative to non-union wage levels. Fig. 4.3 traces the decline during the 1980s of the union relative wage as measured by the Employment Cost Index, which holds occupation and industry constant (US Bureau of Labor Statistics 1991). In more modest post-war recessions, union wages have gained against non-union wages.

During the same period, the form of union wages began to change in an unprecedented manner, as many US unions accepted lump-sum and/or two-tier wage payments. The former are annual payments that are often contingent on the performance of the

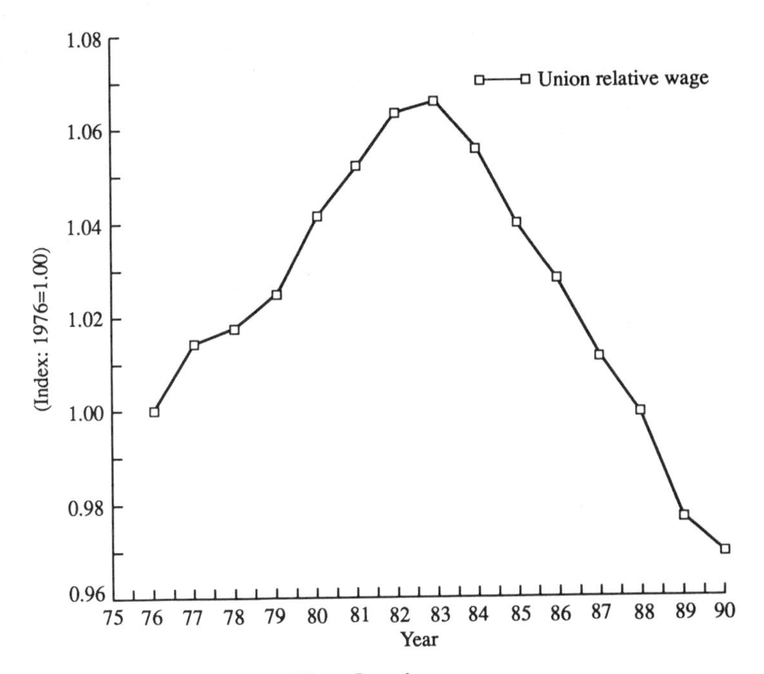

FIG. 4.3 Union Relative Wage Level

firm (and, as discussed above, offer a riskier income stream to some union members), while the latter place new employees on permanently lower wage scales than existing employees.[2]

The US data reveal a particular pattern of adjustment of union contracts to a change in the economic climate. Initially, maintenance of the contractual status quo produces a large quantity adjustment (predicted by many bargaining models). Although delayed, wage adjustments resulting from current bargaining decisions are very rapid once begun and include some shift toward novel (for the union sector) compensation forms indexing pay to the performance of the organization and creating inequality among different classes of union members. (Overall union wage adjustments are none the less somewhat retarded by the inertia of deferred wage increases provided for by multi-year contracts negotiated in earlier years.)

[2] See Flanagan 1990 for further details and implications of US union wage adjustments during the 1980s.

Both the initial emphasis on a quantity adjustment and the speed and forms of the later wage adjustment are consistent with a median voter interpretation of union decision-making. In a world of seniority-based lay-offs, the median union voter is unlikely to support relatively flexible wage policies during the normal range of economic fluctuations. So long as lay-offs reach only a minority of the bargaining unit, flexible wage policies fail because they threaten the income security of more senior workers. Median voter considerations imply that more flexible wage arrangements would only occur when retaining a fixed-wage system implied plant closings and/or layoffs that would threaten the job security of the median union member. Until this threat was perceived in the early 1980s following the massive quantity adjustments described earlier, unions adhered to standard wage arrangements.

A median voter argument also explains why unions, historically committed to producing greater wage equality for their members, would accept contractual arrangements, such as two-tier wage increases, producing distinct inequality. When unions become convinced that survival of an employer requires reduced labour costs, they face two choices. One is an across-the-board reduction in wages that preserves the objective of 'equal pay for equal work'. The other is preserving the wage level of current union members (who must ratify the contract), while shifting the brunt of the wage adjustment to future members (who do not vote on the current contract). With a sufficiently short-run perspective, the two-tier structure wins.

In contrast to the foregoing interpretation of developments in the United States, some versions of the 'insider–outsider' literature have used self-interested voting to rationalize increasing wages in the face of unemployment in Europe (Lindbeck & Snower 1988; Blanchard & Summers 1986). In these models, employed union members ignore the interests of unemployed members when setting wages. With adverse demand shocks, some insiders lose their jobs, but when demand improves, the remaining insiders raise wages with only their continued employment in mind. How can theory reconcile apparently divergent behaviour of unions in the United States and Europe? The key suggested by the median voter theory is in who votes. Under US institutional arrangements, unemployed union members retain their voting rights for a period of time, so that as unemployment grows, the job security of the median voter is effectively threatened at some point. Under European institutional arrangements, unemployed members do not vote, so that policy is determined by the median of employed members.[3]

Despite the intuitive appeal of median voter stories based on an ordering of preferences by seniority, their explanatory power may as a practical matter be limited to unionized labour markets in North America. Elsewhere, work-sharing arrangements rather than lay-offs by seniority appear to be the common pattern of adjusting labour inputs to economic fluctuations. This difference, visible in comparative employment–output elasticities, arises from the greater use of shorter weekly hours or fewer weekly workdays as an adjustment mechanism in Europe and Japan (Grais 1983; Moy & Sorrentino 1981).

This chapter reverses the paradox noted at the end of Chapter 3: multi-dimensional classical voting models resemble voting over union contracts, but imply an instability in union policies that is not observed. In contrast, median voter models provide distinct equilibria that appear descriptive of many outcomes in union labour markets, despite the fact that union settings do not appear to satisfy the prerequisites for the median voter model. The following chapter explores an approach to resolving the lack of congruence.

[3] Empirical tests of the insider–outsider hypothesis have produced inconclusive results so far. Holmlund (1991: 18) reviews some evidence.

5

Institutional Structure and Voting Outcomes

The role of models of unrestricted majority rule in the study of collective choices parallels that of perfect competition in the study of markets. The models isolate forces that are fundamental to understanding the process under study, but are too sparse institutionally to characterize accurately the outcomes of modern realizations of these processes. The contrast between the predicted instability and the observed stability of outcomes signals this tension. After all, risk-averse agents have incentives to modify the uncertainties of unrestricted majority rule by developing internal political processes or norms that channel voting in ways that produce more stable outcomes. We therefore proceed by considering the extent to which various institutional features channel collective choices in unions to produce more stable outcomes.[1]

Two kinds of institutional response in principle could reduce the indeterminacy of classical voting outcomes.[2] First, the development of institutional arrangements that produce greater uniformity of preferences among voting union members should produce comparatively stable outcomes. (As noted, many economic models of union behaviour implicitly assume the presence of such arrangements to the extent that one can consider only the preferences of the 'typical union member'.) Second, the development of internal political processes that reduce the number of issues (dimensions) addressed in membership votes could produce stable voting outcomes. These alternatives are considered in turn.

[1] This notion of 'structure-induced equilibria' was pioneered for legislative settings in Shepsle 1979.

[2] Lindbeck & Weibull (1989) consider a third approach in which voters have preferences for both policy outcomes and for a political party and its candidates. With the addition of some degree of party or candidate preferences, median voter equilibria no longer hold. This particular construction does not correspond to union settings, where parties are rare and the ratification votes are pure policy choices. See, however, the discussion of leadership influence in Chapter 6 below.

Jurisdiction and Preference Homogeneity

The more important sources of preference heterogeneity among union members are age, seniority, skill, the organization of work, and the expected duration of the employment relation. The effects of age and seniority were discussed above. Skill differences produce different preferences regarding wage levels and structure, different organizations of work produce different interests in safety and work rule issues, and differences in the expected duration of the employment relationship produce different objectives regarding the allocation of lay-off risks. The potential heterogeneity is substantial. One influence on the degree of heterogeneity present in actual union votes is jurisdiction—the domain of workers that a union organizes and represents.

An award of jurisdiction accords a union a property right to set the terms and conditions of employment for a particular type of worker, and as such is an institutional arrangement for creating organizational units with some common interests.[3] (The earliest union jurisdictions in many countries were for individual crafts, for example.) With relatively narrow jurisdictions, preferences should be less dispersed and agreement on bargaining objectives should be easier in principle. Preference heterogeneity and hence the risk of voting cycles should grow with the breadth of the jurisdiction (bargaining structure). Other things being equal, one would predict that cycling and instability would be a greater problem in industrial than craft unions and greater still in multi-industrial bargaining arrangements, such as those observed in Scandinavian countries. Nevertheless, this potential is mitigated to some extent by the relationship between bargaining structure and the scope of the agreement, discussed below.

Despite the potential effects of jurisdiction, many issues facing unions remain divisive even in decentralized bargaining systems, in which preferences should be more homogeneous. Wage

[3] In the United States, aversion to rival unionism, in which two or more unions claim the right to organize workers in a particular craft, motivated national unions to establish in 1886 a labour federation, American Federation of Labor (AFL), to serve as a central arbiter of jurisdictional disputes. The assignment of jurisdictional rights was later at the heart of a conflict over whether the AFL should develop industrial as well as craft unions. Ultimately, the federal government placed the determination of jurisdiction in the hands of workers, who select their union representative by majority vote.

increases jeopardize the employment of some workers, and pay equity redistributes money from high-skill to low-skill members. No observer of (for example) disputes during the 1980s over whether or not to accept wage concessions in many industrial unions in the United States would conclude that divergent membership preferences have been eliminated by jurisdiction. This particular institutional arrangement cannot be invoked to salvage the 'typical union member' paradigm.

Bargaining Structure and Voting Equilibria

To what extent does bargaining structure influence voting equilibria? Are union objectives more stable and consistent in countries with centralized bargaining? Two conflicting effects of centralization impede clear predictions on these questions. On the one hand, centralization of bargaining produces greater heterogeneity of membership preferences, and other things being equal, this should increase the difficulty of establishing a stable voting equilibrium. Certainly the notion of a 'typical union member' is least persuasive under centralized bargaining.

On the other hand, centralization also tends to limit the number of potential dimensions of a negotiated agreement. In general, the more centralized the bargaining structure, the more limited the scope of the employment arrangements covered by the agreement. Centralized agreements cannot address effectively the wide variety of labour-relations problems that arise in the many plants, firms, or industries subject to the agreement. Negotiators tend to focus on issues such as wages and hours of work that have a common meaning and method of application across units subject to the agreement, while issues of variable importance across the units may be ignored, despite their importance to some workers. In the limit, centralized bargaining could focus on single issues (e.g. wages), and median voter arguments could be invoked to establish an equilibrium.[4]

In practice, this is probably too hopeful a result. Some centralized bargaining arrangements arise precisely because organized

[4] With widely dispersed preferences, even the formal median voter equilibrium may be transitory, as groups whose ideal points are distant from the median decide that they will be better off under decentralized arrangements.

labour wishes to supplement wage-level goals with other objectives. For example, the centralization of Swedish collective bargaining occurred in part because the LO could not successfully implement a solidaristic wage (pay compression) policy with decentralized bargaining (Flanagan 1987). But then one cannot explain centralization without first explaining the collective preference for narrower wage differentials. Without further institutional detail, the effect of bargaining structure on collective formation of union goals is ambiguous.

Voting Rules

One institutional response to the inherent instability predicted in Chapter 3 would be to develop voting procedures that effectively transform the multi-dimensional setting into a sequence of single-dimension votes. Consider, for example, a union contract-ratification procedure in which the leader presents tentative proposals to a three-person ratification committee (elected representatives of local unions, for example) whose ideal points are again given as 1, 2, and 3 in Fig. 3.2. Assume that the committee is representative of the heterogeneous membership, but recall from Chapter 3 that this assumption provides no guarantee of an equilibrium in a multi-dimensional collective choice setting.

If the union's voting procedures permit the leader to seek ratification on one issue at a time, (s)he will first offer a wage proposal of $W_{m'}$ corresponding to the preferences of representative 3, the median voter on the wage dimension, and then the non-wage proposal of $N_{m'}$ corresponding to the preferences of representative 2, the median voter on this dimension. Under these procedures contract C* will be a stable equilibrium, in contrast to the earlier discussion of Fig. 3.2 in which the entire contract package was considered in one vote. Note that this is an equilibrium in the sense of simple vote-counting. If preference dispersion is greater on (say) the non-wage dimension and N_m is a considerable distance from the preferences of the outlying representatives and their constituencies, one could anticipate demands for decentralized bargaining over job security that would produce greater variation in outcomes on this dimension (e.g. n' and n'').

In the process just described, the elements of a preferred contract are determined in sequence with full attention to the diverse interests of the membership. The total labour cost of the agreement emerges passively as a 'residual' from this decentralized process. Alternatively, a union might invert this process and reduce the choice to a single dimension by having the membership first vote on the total cost of the package. With the cost collectively chosen (by the median voter), the union could then go on to determine the distribution of this cost among the different elements of the contract. (This procedure has been practised in varying degrees by relatively centralized or 'corporatist' trade unions in Austria and in some Scandinavian countries.) We now have two institutional mechanisms for reducing the dimensions of a union contract to permit a median voter equilibrium. The only difference between the two approaches is in the order in which decisions regarding total cost and its distribution are made. An interesting question is whether the order makes any difference to the degree of cost pressure.[5]

Fig. 4.4 replicates the collective-choice environment of Fig. 3.2. Each point in the figure indexes both the cost of the contract and the division of costs between wage and non-wage provisions, and points 1, 2, and 3 represent the 'ideal points' of representatives of different membership constituencies. Consider now the cost that emerges from the 'centralized' determination of union objectives, in which the union first establishes a total labour cost for the contract and then determines the distribution between wage and non-wage items.

Sophisticated union representatives will take account of the likely outcome of the later vote on distribution between wage and non-wage expenditures when initially choosing the labour cost level. Suppose labour cost level LC1 is chosen in the first stage. How well off will the three representatives be in the vote at the second stage? Each representative prefers (Euclidean) indifference curves that are closer to their ideal point. Therefore, the intersection of LC1 with the shortest possible radius from the ideal point to LC1 is the best contract that each representative can achieve in the second stage. That is, the most favourable

[5] Ferejohn and Krehbiel (1987) develop this model in a legislative budget-determination setting.

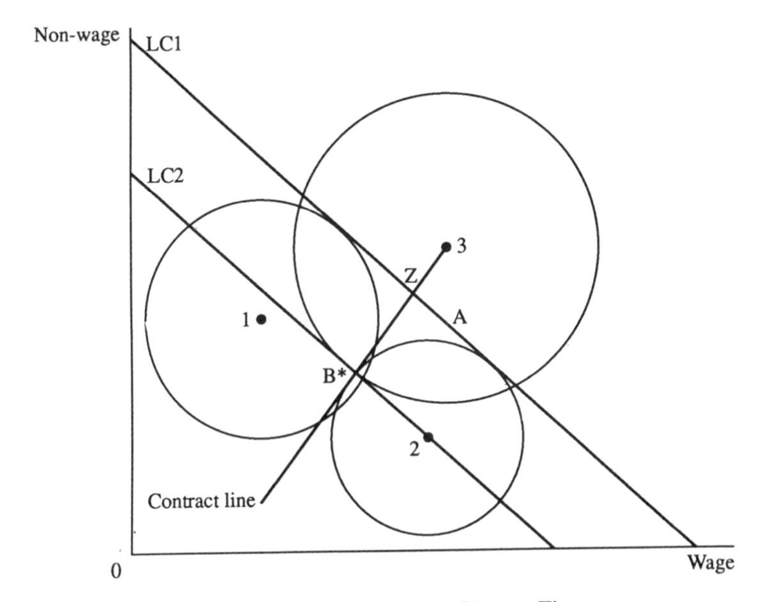

FIG. 4.4 Equilibrium with Contract Cost Chosen First

feasible indifference curve is tangent to the labour cost constraint. The median of the ideal points projected on to the LC1 cost line, Z, represents the contract chosen collectively at the second stage given the labour cost constraint chosen at the first stage. Any other point consistent with that cost level will make a majority of representatives worse off. For different labour costs the collective choice equilibria of the union representatives is therefore given by the 'contract line', which passes through the preferred point of representative 3 and is perpendicular to the labour cost lines. (The 'contract line' in Fig. 4.4 should not be confused with the 'contract curve' of bargaining theory.)

What total cost will the representatives of the various union constituencies be able to agree on? As long as a majority prefers an outcome at a different labour cost, as they would at Z (where a majority consisting of representatives 1 and 2 prefer a lower total labour cost), there can be no equilibrium under the centralized procedure. Given the configuration of preferences in Fig. 4.4, the equilibrium contract will be at B* on LC2, which a majority prefers over any other labour cost equilibrium. (Notice that LC2 is the total cost preferred by representative 2, the median voter on the total cost issue.) In comparison, under the sequential voting procedure analysed earlier, the equilibrium contract would have been A. In Fig. 4.4, therefore, the 'centralized' procedure for formulating bargaining objectives produces lower labour costs than the sequential 'decentralized' procedure.

Unfortunately, this is not a general conclusion; it rests on the particular configuration of preferences in Fig. 4.4. With different preferences, the decentralized procedure could produce lower costs. Fig. 4.5 illustrates the two possibilities in condensed form. The equilibrium labour cost line with contract B* chosen under the centralized procedure can be compared with the sequential result, contract A, that emerges from the median voters on each dimension. Intuitively, centralized determination is only superior when the (minority) preference outlier prefers a high-labour-cost contract. (There is a parallel to the relative position of the metal-workers' union in Sweden prior to the centralization of bargaining (Flanagan 1987)). Without information on the configuration of preferences, analysis of the collective choice process does not predict whether the centralized approach to formulating union objectives yields lower labour cost pressure.

This section added a richer institutional environment to the earlier analyses of unrestricted majority rule to indicate mechanisms for channelling union voting in ways that produce stable equilibria. It also indicated how particular voting procedures can

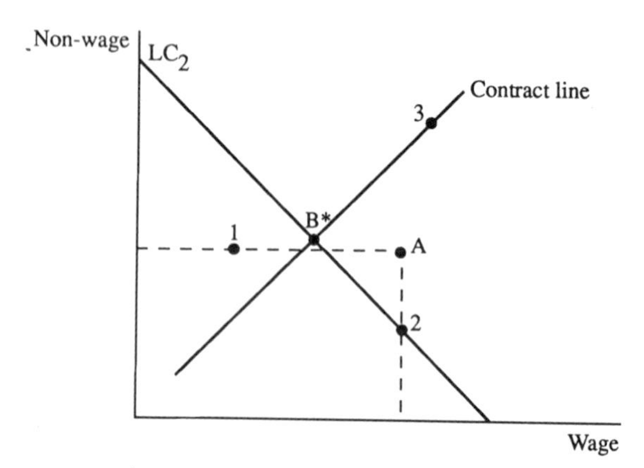

FIG. 4.5 Centralized Voting Produces Lower Labour Costs

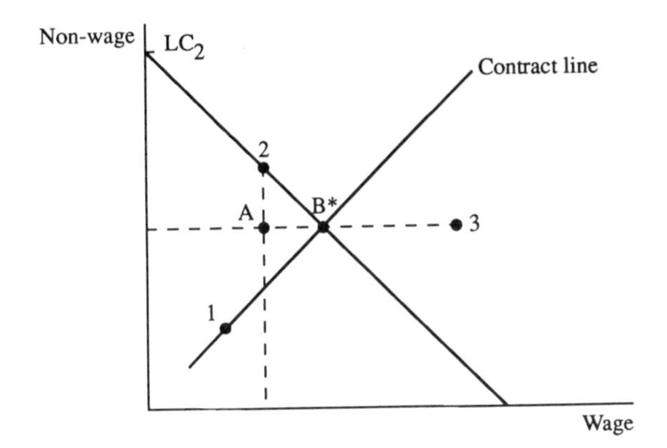

FIG. 4.6 Centralized Voting Produces Higher Labour Costs

influence total labour costs. Like the 'chaos' result in Chapter 3, each of the voting mechanisms examined above rests on particular voting sequences. Yet, proposed labour agreements in many unions (and in virtually all unions in the United States) are ratified by a single vote on the (multi-dimensional) proposal, rather than a sequence of votes on its individual elements.

Ratification by a single vote greatly reduces the potential number of acceptable new contracts. Referring again to Fig. 3.2, with unlimited voting agendas a new contract could be anywhere in contract space, as demonstrated in Chapter 3. With a single ratification vote, the options are reduced to contracts in the cigar-shaped win sets. In reducing the domain of acceptable contracts, a single ratification vote does not produce a distinct equilibrium, but significant departures from the initial contract can only occur gradually over a sequence of (often multi-year) contract negotiations. As will become apparent in Chapter 6, this feature of union ratification votes also limits the scope for independent influence of outcomes by union leaders.

6
Union Leaders and Union Members

Union members delegate considerable authority to their leaders, including the negotiation of labour agreements and the implementation and enforcement of labour contracts. This basic organizational feature receives scant attention in modern economic analyses of union behaviour. By assumption, union leaders faithfully pursue the interests of members, just as in first-generation theories of the firm, managers faithfully pursued the objectives of owners. In fact, delegating such authority exposes union members to the risk that their leaders may substitute personal goals for membership goals. Indeed, judging by the shape of post-war US labour relations legislation, the US Congress and perhaps the public at large seem to assume that this is the rule.

A key question for the analysis of union behaviour is to what extent and under what circumstances union leaders can follow policies that deviate from membership objectives. Any evaluation of the desirability of such deviation is likely to be situation specific. After all, leaders who have the scope of action to squander union funds or pursue other venal objectives may also be able to pursue actions that are in the long-run interest of the union organization although they are resisted by the membership. The issue for this essay is methodological rather than normative, however. If leaders can act independently of members, many of the problems developed in the preceding chapters can be set aside, and attention focused on the leader's objectives. At the same time, the preceding chapters have signalled that the issue is trickier than (say) the question of whether managers act in the owners' interest, since the collective preference that a faithful leader might pursue may be difficult to determine. Ultimately, the issues addressed in this chapter raise the question of why unions have leaders.

Agenda Control

The earlier discussion of classical voting theories stressed the indeterminacy of unrestricted majority rule. Unguided voting can lead almost anywhere, in large measure because voting agendas are formed randomly in this rather spare institutional setting. Viewed from a slightly richer institutional perspective, however, these same theories support a radically different interpretation: the person who determines the agenda controls the outcomes. In principle, anything is possible. In fact, the outcome depends on the exact voting sequence followed.

Consider again the membership preferences in Fig. 3.1. A union leader who personally preferred job security guarantees over a wage increase would establish a two-stage agenda in which members first voted between a pension or a wage increase and then in the second stage between the winner of stage one (pension) and job security guarantees. Notice that job security, which wins the second vote, would have been defeated in a direct contest with wages. A parallel interpretation holds for multi-dimensional, representative voting illustrated earlier in Fig. 3.2. The 'chaos' interpretation stressed in the earlier discussion again rests on random selection of an agenda. In the context of leadership power over members, we now stress the parallel implication that anyone who controls the agenda can in principle establish a sequence of votes that leads from any one contract in Fig. 3.2 to any other contract.

In this setting, union leaders can capture the gains from collective action by manipulating a voting sequence to obtain their objective. With agenda control, outcomes depend only on the preferences of the agenda-setter, which in principle can be quite different from those of the 'average' member. Indeed, agenda control implies that the problems associated with determining the collective preferences of union members can be safely ignored. Unions have leaders in order to produce equilibrium out of the potential chaos of unrestricted majority rule. Thus, agenda control can produce stability of outcomes that depend on the preferences of the agenda-setter, and policy changes are now associated with changes in the identity of the agenda-setter.

The central question raised by agenda-control models is why a union membership (or public policy) would accede to such manipulation by a union leader. This particular device for achieving a stable equilibrium would seem to be purchased at a high price. Raising this question merges discussion of agenda control with the principal–agent considerations that are explored in the following section. Before turning to that discussion, we note procedural and public policy limitations on the effective exercise of agenda control in unions.

Procedural Limitations

The exercise of agenda control can occur in two institutional environments. In one environment, illustrated by the example of agenda control constructed earlier from the preferences in Fig. 3.1, voting on a sequence of pairwise choices permits a union leader to manipulate the sequence to obtain a desired outcome. As a practical matter, this appears to be ruled out by actual union voting procedures. A proposed contract is not presented to union members or their elected representatives as a sequence of pairwise votes; it is presented for approval or disapproval as a (multidimensional) package. Indeed, the absence of voting sequences in unions might be interpreted as an institutional response to limit the potential for agenda control by union leaders and, as noted at the end of the previous chapter, to increase the predictability of acceptable outcomes.[1]

Agenda-control (leadership influence) is also possible when there is only a single vote, if the consequence of rejecting the leader's proposal is reversion to an outcome that is even less appealing to the membership. The more detestable the alternative, the more powerful the leader (Romer & Rosenthal 1979). In a union setting, the notion of a reversion point is a bit fuzzy, but centres on the costs of rejecting a proposed contract—such as continuing to work under the terms of the prior contract or incurring the costs of striking for better terms. In principle, sufficiently high strike costs could confer substantial power on a

[1] The absence of voting sequences also rules out the possibility of 'sophisticated' or strategic voting in unions. Note that voters in the unrestricted majority rule model are surprisingly myopic. At each stage in a voting sequence, they vote strictly in accordance with their preferences, even when it is not in their self-interest to do so. Opportunities for strategic voting—misrepresenting one's true preferences in some votes in order to obtain the most preferred outcome at the end of a voting process—arise where voting sequences exist, and such voting can limit an agenda-setter's power to a certain extent (Farquharson 1969; Shepsle & Weingast 1984).

union leader, particularly since in the nature of the bargaining process, members can only vote the proposal up or down. There is no opportunity for direct amendment.

A very interesting feature of many American union constitutions, however, is that they require larger majorities to authorize a strike than to reject a proposed agreement. Thus, a simple majority can reject a proposed contract, while a strike may require the support of two-thirds to three-quarters of the membership. One effect of this unusual voting rule is to undermine the potential for agenda control by reducing the cost of rejecting a leader's proposal. The default point is cheap, since the leader is simply sent back to negotiate.[2] (In fact, union memberships reject 10 to 15 per cent of tentative agreements in an average year in the United States (Federal Mediation and Conciliation Service).) This particular configuration of voting arrangements effectively provides members with an indirect amendment procedure that limits leadership power.

Public Policy Limitations

The ability of union members to replace leaders who pursue personal rather than organizational goals is a powerful constraint on agenda control, although electoral competition may at times require statutory assistance. Many American unions practised more-or-less unrestricted agenda control prior to 1959. The tenure of office for some union leaders was unusually long by the standards of other voting institutions, and some unions restricted eligibility for elective office or imposed trusteeships enabling the national union to take over the administration of local unions that were a source of political opposition. The leadership controlled discussions at national conventions by the appointment of committees which passed on delegate credentials and developed proposals for consideration on the floor of the convention.[3]

In 1959, following several years of Congressional hearings

[2] An interesting but untested implication of this argument is that leadership power is greater in unions in which contract ratification votes and strike votes require the same majority.

[3] One study of 70 national unions in the late 1950s concluded that 'in the majority of cases, control of the committees is in the hands of the union president, and there is no real rank-and-file check upon his power' (Bromwich 1959: 13). In 49 of the unions, committees were appointed by the president and in another 11 they were appointed by the Executive Board. Only 4 of the 70 constitutions provided for election of committees.

detailing the corrupt and undemocratic practices of a few unions, Congress passed the Labor–Management Reporting and Disclosure Act (LMRDA). The intent and impact of this legislation was to produce a once-and-for-all reduction in agenda control within unions. The law requires elections for national union officers at least once every five years and elections for local union officers at least once every three years. If serious election irregularities occur, the US Secretary of Labor may seek federal court approval to set aside the election results and rerun the election. In several instances, including some of the largest national unions, the Secretary of Labor used the authority of the LMRDA to conduct rerun elections that reversed initial outcomes.[4] It may be significant for the analysis of union behaviour elsewhere that legislative restrictions on agenda control appear to be rare in other countries.

The introduction of agenda control in the sense of manipulation of union members by a leader produces distinct voting equilibria. These equilibria might assist the formal study of union behaviour if the union leader's objectives were clear, but on this issue there is little agreement. In any event, complete leadership control of union objectives requires unnaturally passive behaviour by union members. At the same time, members may wish to cede some autonomy to leaders to produce an effective organization. Thus, a credible theory of agenda-control would have to specify why union members cede some opportunities for manipulation to a union leader and how they simultaneously constrain the leader's freedom of action. Introducing opportunities for members to manipulate leaders stands the initial notion of agenda control on its head and restores both the primacy of membership objectives and the problems associated with collective determination of those preferences. It also invokes the principal–agent problem, which is discussed in the next section.

[4] The LMRDA included many other requirements that facilitated electoral competition in American unions. One indication of the extent of pre-1959 agenda control in unions, comes from follow-up studies by the Department of Labor, which indicate that about three-quarters of the union constitutions were amended in the five years following the passage of the LMRDA and that the most common changes were in provisions specifying terms of office, the requirements for secret ballot elections, adequate notice of elections, and the preservation of election records.

Principal–Agent Interactions

The relationship between union members and leaders fits a standard principal–agent framework. Members (the principals) delegate authority to a leader (the agent) in order to increase the effectiveness of the union organization. (How this might occur is discussed below.) The usual potential for conflict of interest between the personal goals of the leader and the goals of the members exists. In addition, the collective choice setting presents leaders oriented towards serving the union with particular problems. First, the goals of the members are likely to be more ambiguous than in the case of a firm for reasons discussed earlier. Second, a leader who pursues goals that are important for the survival of the union may be in conflict with the objectives that emerge from the union's collective choice process.

Members cannot observe the leader's effort much of the time, but periodically receive information on outcomes, such as tentative negotiated agreements. These outcomes, whether 'good' or 'bad', will reflect the influence of the leader's effort and exogenous events beyond the control of the leader in unknown proportion. Members' efforts to monitor the performance of the leader *ex post* are frustrated by a noisy output signal.

Contrary to the assumption in agenda control models, union members also have tools at their disposal to try to minimize the risk inherent in the principal–agent relationship. They can create incentives *ex ante* for the leader to pursue their objectives, and these incentives can allow for the fact that the leader's actions are only one influence on final outcomes. Theoretical discussions of principal–agent relationships stress that control of the agent requires both a 'compensation rule' and a 'firing rule' relating rewards (punishments) to *ex post* evaluations of performance. Both of these incentive mechanisms exist in unions in the forms of leadership salary arrangements and electoral competition.

Regarding compensation mechanisms, the key issue is whether leadership compensation is related to (relative) performance. Merely raising this issue brings us back to the question of identifying the principal's (membership's) performance objective. As demonstrated earlier, this is far less obvious in a collective choice setting than in the case of a private firm. On the other hand, one could take an empirical approach to this question by studying the determinants of union officer compensation to determine what goals union officers are rewarded for pursuing. To date, this approach to resolving the theoretical indeterminacy of union objectives has received little attention.[5]

There are none the less reasons to doubt that compensation provides a powerful mechanism for controlling union leaders. In a world of two- and three-year contracts, opportunities to monitor leadership performance appear to be limited. Salaries are normally changed every several years, usually by votes taken at union conventions. A few union constitutions contain provisions linking leaders' salaries to the basic wage for members, but this is not the rule. Explicit rules linking compensation to relative performance are apparently unknown, as are salary links to more subtle objectives, such as pay equality and the allocation of employment risks across union members. One might note that the weakness of compensation structures as a tool for controlling principals seems typical of all collective-choice institutions.[6] For example, legislators typically receive a common salary unrelated to degree of effort or success in satisfying the different objectives of their diverse constituencies.

The 'firing rule' for union leaders is implemented by electoral competition. Under US law, for example, union officers serve at the will of the membership, and elections must be held at least every five years. Moreover, as discussed above, the LMRDA reduced the possibilities for agenda control by reducing the cost of mounting electoral competition. As with compensation, very little is known about the extent to which the turnover of union officers is related to their performance. Opportunities for implementing the firing rule are considerably less frequent than in the private sector and only by accident would coincide with specific performance acts. (Again, there is a parallel to the American legislative setting.) Moreover, a leader who cannot (because of constitutional limits of term) or will not stand for re-election is not threatened by the firing rule. Theoretical analysis in a repeat-

[5] Ehrenberg & Goldberg (1977) study the relationship between performance and pay for business agents in construction unions. The present author has a study in progress for a sample of national union leaders.

[6] Doubts have also begun to arise about the power of compensation incentives for influencing managers in the private sector (Jenson & Murphy 1990).

play context shows that a firing rule based on observed (absolute) outcomes can influence the agent's actions when the principals are homogeneous. Control of the leader disappears with the introduction of heterogeneity among principals, however, because the leader-agent has 'both the opportunity and the motive to play off the voters against one another' (Ferejohn 1986: 21).

This discussion began with the view that the key to the relationship between union members and their leaders is in how the members might manipulate their leader(s) by establishing performance- or outcome-based rewards and punishments. The subsequent discussion indicated (tentatively, given the limited empirical research on some issues) that the standard compensation and firing incentives for influencing an agent's behaviour appear to lack power in a union (and more generally, a collective choice) setting. Perhaps because of these limitations, union members also have a right not generally available in the private sector to review and reject major leadership actions (notably through contract ratification and strike votes).[7] Given the apparent weakness of standard incentives but the presence of institutional mechanisms for reviewing individual leadership actions, how much scope of action does the leader-agent have?

If the standard mechanisms for controlling agents operate weakly in a union collective-choice setting, as the preceding discussion suggests, how much power does a union leader have to obtain outcomes that differ from the preferences of a fully informed membership? Political scientists have studied this question extensively in the context of whether legislative committees can achieve outcomes other than those desired by the floor of the legislature. The answer (abstracting from considerable detail reviewed in Krehbiel 1988) is 'rarely, and even then only in circumstances that are not found in union settings'. We have already described the crux of the issue: when a fully informed membership can review and effectively request amendments to specific actions of the leader (by rejecting proposed contracts), only leadership proposals corresponding to members' preferences will survive as desired union policy. Understanding union objectives remains a problem of understanding the

[7] Again, there is a parallel in American legislative settings, where the floor of the legislature must review and approve proposals by committees.

collective preferences of the membership. To be sure, final outcomes may deviate from membership preferences because of the costs of employer resistance, but independent leadership influence is unlikely with a fully informed membership.

7

A Stocktaking

Following a critique of the foundations of the union side of modern union–management bargaining models, the essay first examined a series of models in which union leaders are passive and union members have full information. The analysis showed that the notion of a typical union member rests on rather shaky foundations. Under unrestricted majority rule, the typical member could be almost anyone, but not for very long. There are grounds to be suspicious of these models, however, since the instability they predict is not generally observed.

Later discussion showed that one cannot circumvent the problem of forming a collective membership goal by invoking the power of union leaders, at least under assumptions of symmetrical information. Absence of leadership power leads full circle back to the problem of membership goals. In this setting, leaders must pursue members' objectives, whatever they might be and however they might be formed.

There are other ways out of the classical voting dilemma, however. The paper demonstrates that different voting procedures can produce different outcomes (objectives) including different degrees of union cost pushfulness. Here it is important to distinguish between the effects of voting procedures, as stressed in this essay, and the effects of bargaining structures, reviewed in the other essay in this volume. Only the latter has received much prior attention in the economics literature on unions, and the effects of variations in the former deserve more exploration.

Unions can adopt voting rules that produce voting equilibria. Sequential issue-by-issue voting and votes on total labour costs are two mechanisms that can produce an equilibrium out of chaos. By producing single-issue voting, these procedures allow union goals to be characterized as median voter equilibria. Despite their theoretical appeal, such 'structure induced equilibria' appear to be ruled out (at least in the United States) by the absence of voting sequences in most union organizations. Yet, median voter equilibria appear to have considerable descriptive power in union settings. The model of 'centralized' determination of labour costs may be closest to institutional behaviour in some countries, although it leaves unanswered the question of why centralized procedures are chosen in the first place.

A second way out of the classical voting dilemma may be to consider the influence of union leaders on union objectives in an environment of imperfect information. The very existence of union leaders is difficult to rationalize under perfect information. It is more likely that leaders are desired because they specialize in negotiating and monitoring compliance with collective-bargaining agreements—a specialization that enhances organizational efficiency. At the same time, leaders may exact a price in exchange for acquiring these skills for the union. Union members may have to cede some of their objectives so that the leader has sufficient incentive to specialize in the union's problems. That is, the leader may require some scope to 'lead' in the sense of pursuing some objectives that might not emanate from the membership's collective-choice process. The nature of this particular interaction would seem like a fruitful area for future research.

The essay highlights the role of alternative voting procedures in the formation of collective union preferences. Clearly, the possibilities for (*a*) a stable equilibrium in the case of membership control or (*b*) agenda control by a union leader depend crucially on the specific voting procedures used by a union, so that the way in which union objectives are modelled should vary with the procedures used by the unions under study.

Consider the issue of agenda control. Procedures used by unions in the United States dominate the discussion in this essay, which concludes that given the voting arrangements found in most US unions and public-policy requirements, the scope for agenda control is too limited to play a role in the modelling of union objectives. The focus should be on the development of collective preferences by the membership. A rather different conclusion may be reached under centralized bargaining systems in which the procedures for contract ratification and the initiation of strikes can be quite different. For example, in Sweden and Norway the central union leadership appears to dominate the ratification of collective bargaining agreements and the initiation of strikes (Visser 1987). Subject to the control exerted by electoral competition, this would appear to provide much more scope for leadership influence than one would consider when analysing US union behaviour. Thinking along these lines forces reconsideration of even widespread ideas. For example, can employed insiders implement a selfish wage policy in a union dominated by leaders?

If the political analysis of unions assists conceptual thinking about unions in settings with well-specified voting rules, it does not obviously assist empirical work on union behaviour. The central role of the configuration of membership preferences in determining specific results limits the possibilities for empirical applications without specification of these preferences. The paucity of data on individual voting records in unions precludes testing of preference-based theories that is often possible in legislative settings. On the other hand, one must consider the current alternative—specifications of union objectives representing the assertions of economists.

8
Comment

ALISTAIR ULPH

My comments are in three parts. First, I shall comment on the rationale behind this essay. Second, I shall give a slightly different interpretation of the voting problem, and discuss some new results by Caplin and Nalebuff (1988; 1991); while these shed a somewhat different light on some of the conclusions Flanagan reaches, the general pessimistic conclusion remains. Finally, I shall assess the implications for the structure of unions and union membership.

Rationale

The concern of this essay is with the lack of serious foundations for how to derive a union objective function from the preferences of its individual members, and the implication of the title is that by using political models to provide such a foundation we might be better able to predict what unions might do. My first question is what aspect of union behaviour might such political models allow us to predict better, and do they succeed? Flanagan does not make it completely clear, but let me suggest four possibilities.

(i) The Outcome of Bargaining

If we consider what bargaining theory tells us about the likely outcome of a bargain, it is clear that the theory gives little specific predictions. It indicates how to derive the bargaining frontier (which is where union preferences come in), and how to select a point on the frontier according to the bargaining strengths of the parties, but a very wide range of outcomes can be rationalized by an appropriate choice of union objective function and bargaining strengths. It may not seem unreasonable then to ask whether political models can give us a better fix on union objectives and hence allow tighter predictions. But it is clear from this essay, and from a priori reasoning, that such a hope is forlorn. After all, we are looking for a way of aggregating individual preferences; even if we know how that aggregation is done, we still do not know what the underlying individual preferences are. Thus, if one of Flanagan's suggested methods—the successive use of median voter models on individual issues—is correct, how do we calculate the median voter's preferred outcome on each issue? If it is the preferences of union leaders that count, how are we to assess them?

(ii) Comparative Statics

It can be argued that it is as silly to try to get information on union preferences to allow us to obtain precise predictions on bargained wages and employment as it would be to try to get information on an individual consumer's preferences in order to predict that individual's consumption of, say, beer. What we are concerned with are the comparative static predictions that emerge from embedding these preferences in a model of bargaining in the first place or consumer choice in the second. Thus we want to know how wages or employment are likely to respond to exogenous changes in the nature of the bargaining problem, such as a fall in the demand for the output of a firm with whom a union is bargaining, or the introduction of new technology. Even here it can be argued that bargaining theory gives little guidance; will a fall in product demand lead to a cut in wages, in employment, or both? It is well known that, depending on the labour demand curve, union preferences, and what is bargained over, one can get different answers to this question (see, for example, Ulph & Ulph 1990). Of course, as this suggests, the inability of bargaining theory to give an unambiguous answer to this question arises from factors other than just uncertainty about union preferences. But the objections raised in the previous point about whether political models can throw any additional light on the nature of union preferences, to the extent that these are relevant, still applies.

(iii) Stability of Preferences

The above assumes that preferences of unions, however derived, remain stable over time. Yet, as Flanagan shows, there have been times when union behaviour has changed dramatically, as in the collapse in wages in 1981/2 and the introduction of new forms of wage contracts. Now it may be possible to explain such behaviour with a model in which union preferences remain unchanged, but it seems more likely that such a sharp change in behaviour was a result of a change in union preferences. Since there is no reason to believe there was any change in underlying individual preferences, then it seems that there is a role for a political model to explain why there are changes in the way individual preferences are aggregated into union preferences.

(iv) Union Membership and Structure of Unions

The above discussion follows the usual convention of assuming a given membership of a union whose preferences are to be aggregated. It is well known that conventional bargaining models pay little attention to what might determine union membership, or the number of unions that might exist. Again there may be a role for political models to answer such questions.

It seems to me, therefore, that the role of political modelling is likely to be greatest in explaining points (iii) and (iv) above. Of course this will in turn have an impact on (i) and (ii).

Alternative Interpretation of Voting Problem and New Results

I interpret Flanagan's analysis of voting problems as being that the use of simple majority rule voting predicts endemic voting cycles (the fact that any one proposal can always be overturned by some other proposal in a simple majority vote, and that one can get from any one proposal to any other proposal by a sequence of such pairwise votes), unless we can invoke the median voter model assumptions of single-peaked preferences for individuals and single-issue voting. We observe fairly stable behaviour by unions, with a preference for the status quo, so unions must either be able to overcome the problem of voting cycles by ensuring conditions for the application of median voter models apply, or by being able to moderate the consequence of voting cycles—agenda control by leaders.

I want to argue that the voting problem can be interpreted in a slightly different way, which has acquired added force with new results by Caplin and Nalebuff (1988; 1991), and that this throws a rather different light on some of Flanagan's conclusions.

The different interpretation relates to a point Flanagan notes but does not elaborate on. That is that voting cycles can always be eliminated in practice by the use of *supermajorities* rather than simple majorities; that is that in a vote between one proposal (deemed the status quo) and another proposal, a majority of more than 50 per cent is required to overturn the status quo. Flanagan notes that unions do actually make use of such supermajorities in strike decisions, as a way of reducing agenda control. As a simple illustration of how this overcomes voting cycles consider the use of 100 per cent supermajority, i.e. one needs unanimous approval before overturning any staus quo point. Then it is well known that the set of all Pareto efficient allocations are proposals which cannot be overturned by a unanimous vote (since, by definition, someone always loses), while for any Pareto inefficient allocation there is always some other proposal which can defeat it. So there exist proposals which cannot be defeated under 100 per cent supermajority. The problem now is that there are too many such proposals (Fig. 8.1 shows the Pareto set for Flanagan's example in Fig. 3.2). The theory does not tell us which Pareto efficient point will emerge; but the one which is selected will remain unchallenged as long as preferences remain unchanged.

If 50 per cent majority leads to no proposal emerging as an equilibrium, and 100 per cent majority leads to rather a lot of proposals emerging as equilibria, it is reasonable to ask whether there is some intermediate critical majority at which there are no voting cycles, and the set of proposals is smaller than the Pareto set, ideally unique. Such a majority does exist. The problem is that it is rather difficult to calculate this majority. There are results which give an upper bound for the lowest majority at which no voting cycles emerge; for unrestricted sets of preferences this upper bound is $n/(n+1)$, where n is the dimension of the set of issues amongst which choice is made (in Flanagan's

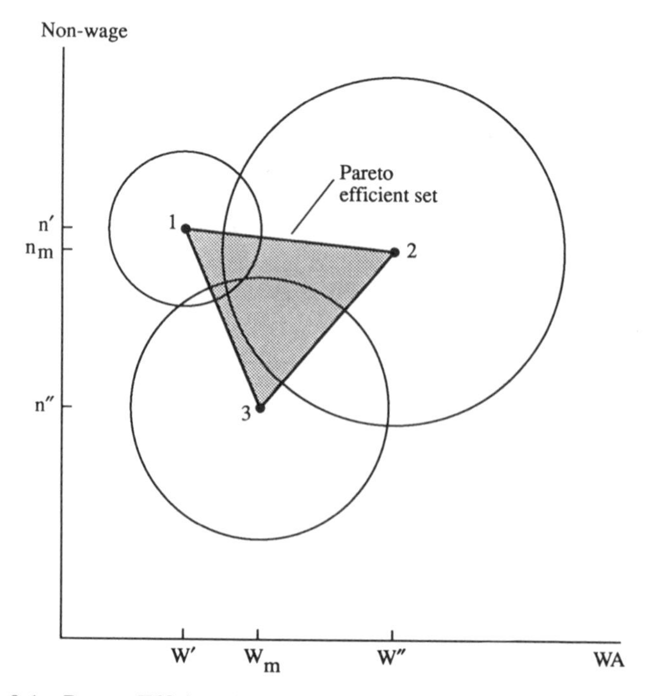

FIG. 8.1 Pareto Efficient Set

example the number of dimensions is 2, wages and non-wages). In practice the number of dimensions could be quite large, so if one was going to set a single majority at which one guarantees that no matter what problem one was faced with voting cycles could not occur, that majority would have to be 100 per cent.

This suggests that there is a rather unappealing choice between risking voting cycles and agenda contol, or requiring unanimity and getting a large set of outcomes with the current status quo point being unchallengeable. However, recent results give slightly more promise. To get round the difficulties one argues that in practice there may be some degree of social consensus among union members such that the kind of preferences they hold, and more importantly, the distribution of such preferences across members, is such that one can eliminate voting cycles at much lower majorities. The work of Caplin and Nalebuff (1988; 1991) gives a set of relatively general conditions on individual preferences (weaker than single-peakedness or their generalization to the kind of Euclidean preferences used in Flanagan's example) and their distribution under which:

(*a*) for any decision problem there exists a critical majority at which there will be a unique equilibrium outcome; if a majority lower than the critical one is set, voting cycles will occur; if a higher majority is set there may be more than one equilibrium proposal;
(*b*) the unique outcome is the one which the *mean voter* would select;
(*c*) the critical majority rises with the dimension of the issues under discussion; the upper bound on the critical majority as the number of dimensions gets very large is 64 per cent.

There are a number of implications emerging from these results that relate to what Flanagan says in his essay. First, it is not necessary to restrict oneself to single issue decisions in order that voting will produce a unique outcome. So the fact that unions are observed to vote on multi-dimensional issues need not be a problem. Second, the fact that the unique outcome corresponds to the choice of the mean voter may give a foundation for the use of union objective functions that are taken to reflect the preferences of the 'typical member'; I do not think it is necessary to follow Flanagan's interpretation of such models as requiring an assumption that all union members are identical.

The third implication is that in practice it is clearly unrealistic to think of unions setting critical majorities every time a vote is required, in the absence of any simple objective rule for determining such a majority. Unions will do what they are observed to do—set a small number of critical majorities according to broad classes of situation. This might give some basis for a theory of why union preferences might seem to undergo radical change. Suppose for a particular class of problems one got precisely the right critical majority at some point of time, say 60 per cent. Using that majority would select the mean voter's optimal choice, point *A* in Fig. 8.2. In subsequent votes on the same class of problems, the use of the same 60 per cent majority turns out to be rather too high; so that although at some later date the mean voter's preferred point has changed to *B* in Fig. 8.2, there is a set of proposals around this point which cannot be overturned by a 60 per cent majority; that set includes point *A*. Over time prefer-

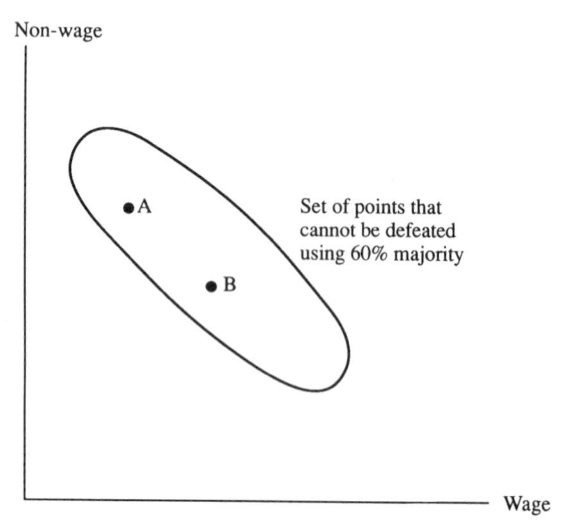

FIG. 8.2 Status Quo Undefeated

ences continue to change in a fairly steady way, so that at some third date the mean voter's preferred point has changed to point *C* in Fig. 8.3. Perhaps more importantly, the number of dimensions by which point *C* differs from point *A* may now be quite large (obviously I cannot show this in two dimensions), so that the set of points around point *C* which can be defended by a 60 per cent majority in this dimension of issues is now quite small. In any event, by a combination of further drift in preferences and some shrinkage of the set of unchallengeable proposals, point *A* now lies outside the set of points around point *C*. Now there will be at least 60 per cent of union members who prefer point *C* to point *A*. So we have a story which explains why one might see rather prolonged periods in which the status quo is defended despite changing circumstances, followed by a sudden, and quite radical (i.e. high dimension) change. This would be consistent with the events described by Flanagan in the United States in the 1980s.

Unfortunately there is a problem with this last story. At least on the evidence for the United States produced by Flanagan, it appears that for most issues only a simple majority is used, and with a simple majority and more than one dimension, then as

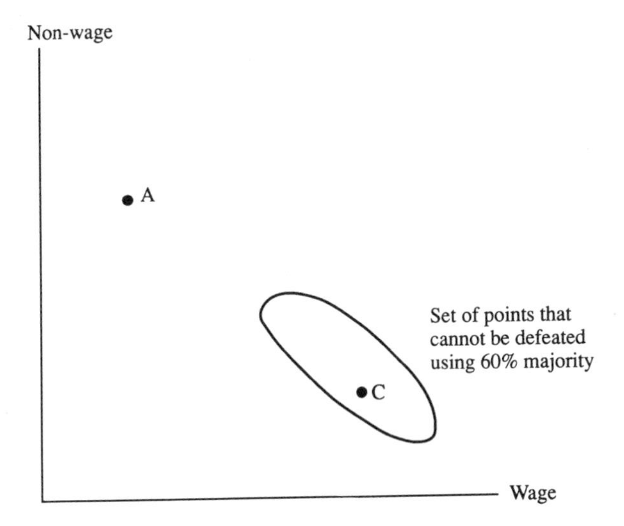

FIG. 8.3 Status Quo Defeated

Flanagan shows, voting cycles are likely to be endemic, even if on any single dimension a unique decision could be selected. This raises the interesting question of why unions appear not to make more use of majorities greater than 50 per cent to give some protection against voting cycles. Of course the above theory, like the theory exposited in Flanagan's essay, assumes that all members of the union vote. Recognizing that not all members will vote, *if* one assumed that a larger proportion of supporters of the status quo were likely to vote than supporters of a new position, then one would want to reduce the correct critical supermajority in order to provide less protection for the current status quo, so 50 per cent may not be as open to voting cycles as would be the case when all members vote. But I can think of no basis for such an assumption.

Implications for Union Membership and Bargaining Structure

The work of Caplin and Nalebuff just discussed requires an assumption about the distribution of preferences amongst union members to ensure that preferences are not 'too different', which, as I have already indicated, falls well short of the requirement that union members have identical preferences. Flanagan talks about this with respect to issues such as the jurisdiction of a union, and the question of centralized/decentralized bargaining. In trying to model such issues it seems to me to be important to remember that we are not just trying to solve a preference aggregation problem; the aggregated preferences are an input to a bargaining problem. It would also be desirable if the theory could help resolve the question of what determines union membership; i.e. we would like a theory which explains why it is rational for individuals to join a union, which union they join, and why it may be rational for existing union members to exclude others from joining. Before elaborating these points it is perhaps worth noting that this is another area where useful parallels could be drawn from the political literature, in particular the literature on fiscal federalism, Tiebout models, etc.

On the first point, what I have in mind is the simple point that if the implication of the preference aggregation problem is taken to be that we may need to have a number of small unions with relatively homogeneous preferences rather than a few large unions with diverse preferences, then one needs to consider the trade-off between the costs of solving the preference aggregation problem and the costs of dissipating bargaining strength. Now it may be that there is no such trade-off, or rather the loss of bargaining power from having separate unions is less than is sometimes supposed, and I note two pieces of work which suggest this. If we differentiate workers by skill, the work of Horn and Wolinsky (1988) shows that if workers are substitutes, then organizing in a single union will give them the best outcome in the subsequent bargain with an employer (essentially because this prevents dilution of monopoly power), whereas if workers are complements, then they should organize in separate unions (to get several 'bites at the cherry'). If we interpret 'substitutability' and 'complementarity' in skills as 'similarity' and 'difference' in preferences, then this bargaining argument would seem to reinforce the preference aggregation argument for having workers with relatively homogeneous preferences in the same union. In some work I did (Ulph 1989), I noted that if one also includes the problem faced by workers of being unable to commit to long-run contracts on wages and employment, so that wage-bargaining may reduce the incentives of firms to invest in capital (Grout 1984), then it may pay workers (even if they are identical in terms of productivity) to form into separate unions, because by weakening their bargaining power they may reduce the disincentive for firms to underinvest, and indeed can induce overinvestment by firms to provide them with sufficient capacity to be able to play off one union against another. However, while there are some circumstances where it pays to have a number of separate unions bargaining with employers, this is clearly not always the case, so there is a need to consider the trade-off indicated above. This suggests that the interaction between the preference aggregation problem and the bargaining problem may provide a fruitful explanation of what determines the number of unions in a particular industry.

The discussion in the previous paragraph about what determines the number and size of unions is at a rather macro level. Similar issues arise though if we consider a micro view in which union membership is determined by the decisions of individuals whether to join a union, and of union members as to whether to allow entry. It is well known that there is a dearth of literature providing micro foundations for union membership. Most of the models that do exist are based on distinguishing union members by seniority and then applying the median voter model to determine the outcome of the wage bargain. Almost all of these models have some unsatisfactory features; at the most extreme there is the 'vanishing union' problem, but even models that get round that (e.g. Grossman 1983) have unresolved difficulties (see again Ulph & Ulph 1990 for further discussion). As Booth and Ulph (1988) note, a crucial reason why such models are unsatisfactory is that they do not provide an important role for larger membership to improve the bargaining position of the union. As they show, once one builds a theory of membership using a bargaining model where membership plays a key role (essentially by assuming that a larger union membership depletes the pool of non-union members who provide the firm's outside option) then most of the unsatisfactory features of the previous models can be overcome, and one gets quite a rich theory of union membership. It would be interesting to extend that work to cover a wider set of asymmetries between workers. In the open-shop version of the Booth and Ulph model to determine union membership it is necessary to introduce costs of forming a union, and these are just taken to be increasing in union size. With more asymmetries in worker characteristics one may want to have costs that reflect the difficulty of the resulting preference aggregation problem.

The message of this section of my comments then is that I think a very fruitful line of future research lies in the careful integration of the kind of preference aggregation problems discussed in Flanagan's essay with a theory of bargaining to produce a fully endogenous explanation of the number and type of unions, who decides to join which union, and how existing union members behave.

Conclusions

In conclusion, Flanagan's paper does start to address a neglected aspect of research on union behaviour, and I think it will be useful to explore some of the implications of more recent work on voting models and preference aggregation. I remain doubtful of whether this line of research will actually help much in providing sharper predictions from bargaining theory about either the levels or comparative static properties of wages or employment that emerge from bargains between unions and firms, since the underlying individual preferences will remain unexplained. Where I do see more promising lines of research is in trying to explain, given some distribution of individual preferences, the number, size, and composition of unions that might emerge. Such a theory will not depend solely on the nature of the preference aggregation problem but its integration into a theory of bargaining.

9
Comment

ASSAR LINDBECK

Robert Flanagan has written a thoughtful essay on the influence of the preferences of union members on union demands in bargaining. The relevance and importance of the issue can hardly be doubted. The fact that the essay is characterized by interesting reflections rather than unambiguous results is certainly a consequence of the complexity of the issue itself rather than being a weakness of Flanagan's analysis. However, because of this complexity, it is difficult to see how economists applying union models to the analysis of wage-formation should modify their analyses in the light of Flanagan's discussion.

A starting-point of Flanagan's analysis is that while voting models on a multi-dimensional agenda predict 'cycling' (or 'chaos'), actual demands by unions in the real world are quite 'stable'. However, is the last part of the assertion necessarily correct? Do we not observe that unions, in fact, change their agendas continuously? Some years a union may emphasize the overall wage rate, at other occasions it is the distribution of wages, the level of pensions, the right to participate in decision-making within firms, demands for a reduction in the number of working hours, safety at work, etc. Is it obvious that this shifting emphasis cannot be interpreted just as 'cycling'?

I also wonder whether Flanagan does not over-emphasize the median voter theorem in his discussion. It has been shown by various authors in the theoretical literature on voting that the outcome is quite different if voters have both 'candidate (or party) preferences' and 'issue preferences'. In this case, it can be shown, for candidates who maximize votes (or plurality), that voting converges to a utilitarian equilibrium rather than to a median voter equilibrium (Hinich 1984; Ledyard 1984; Lindbeck & Weibull 1989).

This result also holds if the candidates themselves have issue preferences, and therefore do not just maximize votes, provided the candidates are equally popular. But if one candidate is more popular than the other, a non-utilitarian equilibrium emerges, in which the more popular candidate wins on a policy that lies between his most preferred policy and the utilitarian optimum (Lindbeck & Weibull 1989).

Flanagan argues, however, that these types of model do not correspond to a union setting where party competition is rare and the ratification of proposed contracts is a pure policy choice, rather than a choice of candidates. However, models of this type do not require political parties. It is enough that voters have preferences concerning the candidates (such as their personalities or ideologies), and such preferences are, most likely, relevant when union members vote for the representative assemblies that often have the power to ratify preliminary bargaining agreements. Thus, I am not convinced that these alternatives to median voter models are irrelevant for the analysis of decision-making in unions.

The overall impression of Flanagan's essay is the need for more good empirical research on union behaviour. This holds also for the issue of the autonomy of union leaders in relation to the preferences of union members—another question discussed by Flanagan. Most likely, the conformity of leadership behaviour to member preferences varies considerably depending on the issue under consideration; we may also expect large differences between unions and between countries.

References for Part I

Agell, J., and Lommerud, K. E. (1990), 'Union Egalitarianism as Income Insurance,' Working Paper no. 79, Trade Union Institute for Economic Research, Stockholm.

Arrow, K. J. (1951), *Social Choice and Individual Values*. New Haven, Conn.: Yale University Press.

Black, D. (1958), *The Theory of Committees and Elections*. London: Cambridge University Press.

Blanchard, O., and Summers, L. (1986), 'Hysteresis and the European Unemployment Problem', in S. Fischer (ed.), *NBER Macroeconomics Annual*, Cambridge, Mass.: MIT Press.

Booth, A. (1984), 'A Public Choice Model of Trade Union Behaviour and Membership', *Economic Journal*, 94: 883–98.

—— and Ulph, D. (1988), 'Union Wages and Employment with Endogenous Membership'. Mimeo.: University of Bristol.

Bromwich, L. (1959), *Union Constitutions*. New York: Fund For the Republic.

Caplin, A., and Nalebuff, B. (1988), 'On 64% Majority Rule', *Econometrica*, 56: 787–815.

—— —— (1991), 'Aggregation and Social Choice: A Mean Voter Theorem', *Econometrica*, 59: 1–24.

Ehrenberg, R., and Goldberg, S. (1977), 'Officer Performance and Compensation in Local Building Trades Unions', *Industrial and Labor Relations Review*, 30: 188–96.

Farber, H. S. (1986), 'The Analysis of Union Behaviour', in O. Ashenfelter and R. Layard (eds.), *Handbook of Labor Economics*. Amsterdam: North-Holland.

Farquharson, R. (1969), *Theory of Voting*. New Haven, Conn.: Yale University Press.

Ferejohn, J. (1986), 'Incumbent Performance in Office', *Public Choice*, 50: 1–26.

—— and Krehbiel, K. (1987), 'The Budget Process and the Size of the Budget', *American Journal of Political Science*, 31: 296–320.

Flanagan, R. J. (1984), 'Wage Concessions and Long-Term Union Wage Flexibility', *Brookings Papers on Economic Activity*, 1.

—— (1987), 'Equality and Efficiency in Swedish Labor Markets', in B. Bosworth and A. Rivlin (eds.), *The Swedish Economy*. Washington, DC: Brookings Institution.

Freeman, R. B., and Medoff, J. L. (1984), *What Do Unions Do?* New York: Basic Books.

Grais, B. (1983), *Layoffs and Short-Time Working in Selected OECD Countries*. Paris: OECD.

Grossman, G. (1983), 'Union Wages, Seniority and Unemployment', *American Economic Review*, 73: 277–90.

Grout, P. (1984), 'Investment and Wages in the Absence of Legally Binding Contracts: A Nash Bargaining Approach', *Econometrica*, 52: 449–60.

Hinich, M. J. (1984), 'Policy Formation in a Representative Democracy'. Mimeo.

Holmlund, B. (1989), 'Wages and Employment in Unionized Economies: Theory and Evidence', in B. Holmlund, K. Lofgren, and L. Engstrom, *Trade Unions, Employment, and Unemployment Duration*. Oxford: Clarendon Press.

—— (1991), 'Unemployment Persistence and Insider–Outsider Forces in Wage Determination'. Working Paper no. 92, OECD Department of Economics and Statistics.

Horn, H., and Wolinsky, A. (1988), 'Worker Substitutability and Patterns of Unionisation', *Economic Journal*, 98: 484–97.

Jenson, M., and Murphy, K. J. (1990), 'Performance Pay and Top-Management Incentives', *Journal of Political Economy*, 98: 225–64.

Krehbiel, K. (1988), 'Spatial Models of Legislative Choice', *Legislative Studies Quarterly*, 13: 259–319.

Ledyard, J. (1984), 'The Pure Theory of Two-Candidate Election', *Public Choice*, 44.

Lindbeck, A., and Snower, D. (1989), *The Insider–Outsider Theory of Employment and Unemployment*. Cambridge, Mass.: MIT Press.

—— and Weibull, J. W. (1989), 'Political Equilibrium in Representative Democracy'. Seminar Paper no. 426, Institute for International Economic Studies, University of Stockholm.

McDonald, I. M., and Solow, R. M. (1981), 'Wage Bargaining and Employment', *American Economic Review*, 71: 896–908.

McKelvey, R. D. (1976), 'Intransitivities in Multidimensional Voting Models and Some Implications for Agenda Control', *Journal of Economic Theory*, 12: 472–82.

Moy, J., and Sorrentino, C. (1981), 'Unemployment, Labor Force Trends, and Layoff Practices in 10 Countries', *Monthly Labor Review*, 104: 3–13.

Oswald, A. J. (1985), 'The Economic Theory of Trade Unions: An Introductory Survey', *Scandinavian Journal of Economics*, 87: 160–93.

—— (1987), 'Efficient Contracts are on the Labor Demand Curve:

Theory and Facts'. Discussion Paper no. 284, London School of Economics, Centre for Labour Economics.
Pencavel, J. (1985), 'Wages and Employment Under Trade Unionism: Microeconomic Models and Macroeconomic Applications', *Scandinavian Journal of Economics*, 87: 197–225.
—— (1991), *Labor Markets Under Trade Unions*. Cambridge, Mass.: Basil Blackwell.
Romer, T., and Rosenthal, H. (1979), 'Bureaucrats vs. voters: On the Political Economy of Resource Allocation by Direct Democracy', *Quarterly Journal of Economics*, 93: 563–87.
Shepsle, K. A. (1979), 'Institutional Arrangements and Equilibrium In Multidimensional Voting Models', *American Journal of Political Science*, 23: 27–59.
—— and Weingast, B. R. (1984), 'Uncovered Sets and Sophisticated Voting Outcomes with Implications for Agenda Institutions', *American Journal of Political Science*, 28: 49–74.
Ulph, A. (1989), 'The Incentives to Make Commitments in Wage Bargaining', *Review of Economic Studies*, 56: 449–66.
—— and Ulph, D. (1990), 'Union Bargaining: A Survey of Recent Work', in D. Sapsford and Z. Tzannatos (eds.), *Current Issues in Labour Economics*. London: Macmillan. 86–125.
US Bureau of Labor Statistics (1991), *Current Wage Developments*. Washington, DC.
Visser, J. (1987), 'In Search of Inclusive Unionism' Ph.D. dissertation, University of Amsterdam.

PART II

Bargaining Structure and Economic Performance

10

Introduction

The structure of collective bargaining differs dramatically among advanced industrial societies. In Japan, most organized workers belong to company unions. If American employers had been successful in the 1920s, most organized workers in the United States would also belong to company unions. Instead, American unions in the private sector are organized along a mixture of craft and industrial lines with wages usually but not always set at the firm level. In Germany, industrial relations are dominated on the union side by sixteen industrial unions with jurisdiction over blue-collar workers, white-collar workers, and even Civil Servants within their sector. Wage-bargaining for each broad industry occurs primarily at the regional level. In one way wage-bargaining has been even more centralized in the Nordic countries of Finland, Sweden, and Norway for most of the post-war period. Centralized wage agreements negotiated by the national confederations of unions and employers have typically covered all private-sector workers at the national level. In another way bargaining in the Nordic countries is less centralized in that blue-collar, white-collar, and professional union confederations bargain separately.

In recent years, economists have begun to recognize that such differences in the structure of bargaining may have important effects on the outcome of labour negotiations and on aggregate economic performance. Indeed, the extent to which bargaining occurs by craft, by firm, by industry, or at the national level and the consequences that follow has grown from a topic that concerned mostly specialists in comparative industrial relations to

We draw heavily on the research projects 'Wage Formation and Unemployment' at the Center for Research in Economics and Business Administration, Department of Economics, Oslo and 'Comparative Institutions of Wage Bargaining' at the Institute for Industrial Relations, UCLA, Los Angeles. Most of the paper was written while K. Moene was visiting UCLA. We thank David Ellison for bibliographic assistance. We have benefited from comments by Asbjørn Rødseth and the participants at the FIEF seminar in Stockholm.

become an important issue in economics, political science, and sociology during the past fifteen years. The main reason for this increased interest in the structure of bargaining among scholars outside the Nordic countries was the challenge posed by the divergence of macro-economic performance among advanced industrial societies since the mid-1970s. As different countries responded to the same external shocks with very different combinations of unemployment, inflation, and real wage reductions, much research has focused on national differences in the institutional structure of collective bargaining, in particular on the centralization of bargaining.

Nordic scholars have an additional reason to investigate the economic impact of different bargaining structures. Not since the 1930s has the structure of bargaining in the Nordic countries been in such flux. The systems of highly centralized bargaining that have dominated wage-setting in Sweden, Norway, and Finland since the Second World War have come under great pressure to decentralize in the 1980s. The biggest change has occurred in Sweden where representatives of Swedish employers in the SAF have underscored their opposition to centralized wage-setting by dismantling their capacity to bargain at the national level (Myrdal 1991). In all of the Nordic countries, the increased utilization of profit-sharing and other incentive schemes in compensation packages as well as the greater importance of locally bargained wage drift as a share of total wage growth have raised questions concerning the ability of central bargainers to control wage growth (Elvander 1988; 1989). Thus, the question of whether unions and employers should seek to rebuild the post-war centralized bargaining system or encourage the present trend towards greater decentralization is of immediate concern to unions and employers in the Nordic region.

Answers to the question of how the structure of bargaining affects economic performance have proven to be highly controversial. On the one hand, economists and, increasingly, policy-makers share a belief in the superiority of decentralized price-determination over all forms of centralized price-setting, whether by governments or by collective bargainers. In labour markets, as in other markets, competition and price- (i.e. wage-) flexibility are considered to be good things. To the extent that centralized bargaining reduces competition among workers and diminishes the sensitivity of wages to local conditions of demand and supply in the labour market, the argument runs, economic performance is impaired.

On the other hand, extensive cross-national research has linked centralized bargaining with superior aggregate economic performance along a variety of dimensions. The centralization of bargaining first appeared as an explanatory variable in studies of strike frequency. As early as 1960, Arthur Ross and P. T. Hartman observed that 'the union structure most conducive to the elimination of industrial conflict is a unified national movement with strongly centralized control' (1960: 66). More recent studies by Douglas Hibbs (1978), Walter Korpi and Michael Shalev (1980), and Martin Paldam and Peder Pedersen (1984), among others, have reproduced Ross and Hartman's finding.

Associated with low strike rates is a willingness to co-operate with voluntary incomes policies. Bruce Headey (1970) was the first to demonstrate the association between centralized bargaining and the successful implementation of voluntary wage controls. After surveying all instances of voluntary incomes policies among thirteen Western democracies since the Second World War, Headey concluded that union co-operation was contingent upon two factors: (1) the participation of Left parties in government, and (2) the centralization of wage-bargaining. More recent work by Gary Marks (1986) using a larger set of countries over a longer period of time reaches the same conclusion.

By the 1980s, some index of the centralization of bargaining, either standing alone or as a component of a broader index of something called 'corporatism', was being widely used in studying cross-national differences among OECD countries in the responsiveness of wages to rising unemployment and slowing growth since 1974.[1] The basic argument underlying most of this research

[1] Corporatism is a label much used in political science that eludes rigorous definition. Philippe Schmitter (1974; 1977), who may be credited with introducing the concept in modern social science, defined corporatism as a system of interest representation that is dominated by a small number of encompassing, vertically integrated, centralized organizations. Others, such a Gerhard Lehmbruch (1977; 1979), define corporatism in terms of an intermingling of private and public realms with public policies being negotiated with private interest groups and private decisions being subject to the intervention of public authorities. Yet others, such as Peter Katzenstein (1985) or Newell and Symons (1987), include a commitment by the unions to social harmony and co-operation with employers as part of the definition.

is that the benefits of wage moderation are public goods to an important extent. In the words of the OECD: 'unless wage bargaining is highly centralized, individual unions can rationally hope that an improvement in their real wages can be achieved at the expense of profits and hence employment elsewhere in the economy' (1977: 159). Therefore, centralized bargaining moderates union wage demands. In turn, real wage moderation is widely viewed as the key for regaining low rates of unemployment and inflation and high rates of investment and growth. Thus centralized bargaining is associated with superior economic performance, usually measured in terms of unemployment and inflation. Although the details differ, this is the basic conclusion of numerous empirical studies: Mark Lutz (1981), Wolfgang Blaas (1982), David Cameron (1984), Michael Bruno and Jeffrey Sachs (1985), John McCallum (1983; 1986), Ezio Tarantelli (1986), Charles Bean, Richard Layard, and Stephen Nickell (1986), A. Newell and J. S. V. Symons (1987), Carlo Dell'Aringa and Manuela Lodivici (1990), Richard Jackman (1990), Richard Jackman, Christopher Pissarides, and Savvas Savouri (1990), and David Soskice (1990), among others. Thus either the conventional economic wisdom in favour of decentralized wage-setting is wrong when applied to unionized labour markets, or the empirical studies are flawed.

Indeed, these empirical claims regarding the superiority of centralized bargaining have been challenged. (Only the relationship between centralization and the frequency of industrial conflict remains uncontroversial.) Bernhard Heitger (1987) suggested that the macro-economic benefits of wage-restraint are more than offset by the micro-economic costs of rigid (and, in Heitger's view, overly egalitarian) relative wages that centralized wage-setting produces. Lars Calmfors and John Driffill (1988) and Richard Freeman (1988) have argued that the relationship between centralization and economic performance is hump-shaped rather than monotonic. In their view, countries with both very decentralized wage-setting and highly centralized wage-setting have done better than those in an intermediate position. Peter Lange and Geoffrey Garrett (1985), Garrett and Lange (1986), Alexander Hicks (1988), and Michael Alvarez, Garrett, and Lange (1991) find that countries with centralized unions and social democratic governments, as well as countries with decentralized unions and conservative governments, have done relatively better than countries with one but not the other. Göran Therborn (1987) goes further in arguing that only social democratic governance matters as far as unemployment is concerned. The empirical association of centralized bargaining and low unemployment is a spurious result, according to Therborn, of the high correlation of centralized bargaining and social democratic governance.

Moreover, all of the empirical studies suffer from a number of difficulties. The most notable problem comes from the measurement of the key independent variable: union centralization. While there is consensus that wage-setting in Austria, Norway, and Sweden is (or was in the case of Sweden) highly centralized while bargaining in the United States and Canada is decentralized, many countries are ranked quite differently in different studies. Switzerland, to cite one example, is judged as highly centralized by Bruno and Sachs (1985), moderately centralized by Headey (1970), and very decentralized by Calmfors and Driffill (1988). The outstanding economic performance of Japan, with its system of enterprise unions, is often displayed as presenting strong evidence in favour of the advantages of decentralization. Yet Tarantelli (1986), G. Brunello and S. Wadhwani (1989), and Soskice (1990) claim that wage-setting in Japan is closely synchronized and even centralized in an informal way. A different problem is that few of the studies control for the influence of unions over wages. It seems more appropriate to view the United States in the 1980s as an example of a competitive labour market rather than as a case of decentralized bargaining (Paloheimo 1990). The set of advanced industrial societies is small enough so that removing or reclassifying a few cases can alter the qualitative conclusions.

Given the small number of cases and the large number of factors that plausibly affect economic performance, the credibility of empirical evidence on the advantages or disadvantages of centralized wage-setting depends on the strength of the theory explaining the results. Thus, we have chosen to concentrate in this review on what economic theory has to say.[2] We start, in

[2] See Pohjola 1989 and Tyrväinen 1989 for complementary reviews of the empirical and theoretical literature on centralized bargaining and economic performance. The recent book by Richard Layard, Stephen Nickell, and Richard Jackman contains both theoretical and empirical work that is highly relevant for our topic. Unfortunately we are unable to comment on their work in this review since our paper was completed before the book was published.

Chapter 11, with a review of what can be learned about the impact of bargaining structure from models in which the union is assumed to be able to set wages as it chooses, subject to the constraint that the level of employment (or investment) is chosen by employers. Such models are really models of union aspirations rather than bargaining outcomes. Thus, in Chapter 11 we focus attention on how centralization might affect the militancy or moderation of the unions' wage demands. In section (a), we review briefly the literature on union objectives in bargaining and present the standard, simple model of union wage-setting. In the simplest model with no externalities, exogenous prices, and a single type of labour, centralization has no effect on the unions' optimal wage. In the remainder of Chapter 11, we show how altering these assumptions changes that conclusion. In section (b) we allow wages to affect consumer prices. In section (c) we consider the case with multiple types of labour. Various externalities stemming from union concerns with relative wages and aggregate unemployment are the subject of section (d). The impact of the level of centralization on union preferences regarding the trade-off between wages and employment is discussed in section (e). We end Chapter 11 by considering the question of centralized versus decentralized wage-setting from the point of view of employers when the quality of labour depends on the wages that are paid.

In Chapter 12 we shift from models of the unions' (or employers') optimal wage to models of bargaining. Here we study ways in which centralization affects wage-bargaining holding union aspirations constant. We begin in section (a) with a brief overview of co-operative and non-cooperative bargaining models. In (b) we study the actual degree of centralization that exists in mixed systems of the Nordic variety where centralized wage-setting is followed by supplementary bargaining at the local level. In section (c) we discuss the often noticed but little studied effect of centralization on the frequency of industrial conflict. The subject of sections (d) and (e) is wage-bargaining and profit-sharing. The essential insight is that bargaining at the firm level constitutes a type of profit-sharing that is absent from bargaining at the industry or national level, and that profit-sharing differs from fixed wage contracts in a variety of ways. Profit-sharing has effects on workers' willingness to expend effort on the job (section (d)) and the firms' willingness to hire more workers (section (e)). In section (f) we relax the assumption that the number of firms is fixed in order to analyse the effects of centralized and local bargaining on entry and exit of firms, or, equivalently, on the building of new plants and the shutting down of older ones.

In Chapter 13 we turn from the economic to the political consequences of centralized bargaining systems. By political consequences, we mean the types of conflicts engendered by centralized wage-setting within the union movement and the employers' associations. Such conflicts matter because they threaten the long-run sustainability of centralized wage-setting institutions. Chapter 14 concludes the essay.

11

Wage Demands by Unions and Employers

(a) Models of Union Behaviour

In order to study theoretically how the bargaining structure affects union wage demands, we need to say something about the unions' objectives in collective bargaining.[1] Unfortunately, there is no consensus regarding the appropriate maximand for unions comparable to the standard assumption of profit maximization for firms. Many answers to the question of what unions maximize have been suggested, including wages, aggregate rents, a general function of wages and employment, the utility of the decisive union voters, union dues, and the salary of top union leaders among others. The most basic question, however, is whether union behaviour is consistent with any coherent aggregate preference ordering. On the one hand, to model unions as organizations that seek to maximize union dues or leaders' salaries subject to the constraint that union members will quit if the costs of membership exceed the benefits neglects the real impact of internal democracy on union behaviour. On the other hand, the theoretical literature on voting has demonstrated that the outcome of elections is almost never equivalent to the maximization of some aggregate objective function when heterogeneous voters face choices along more than one dimension.[2] Thus the microfoundations for modelling a democratic union as a unitary actor whose behaviour can be studied as the solution of an optimization problem are easily challenged.

Nevertheless, it is essential in theoretical work that unions be assumed to maximize something. Moreover, the status of theories of union behaviour is not really so different from the theory

[1] See Farber (1986) for a relatively recent review of the literature on union objectives, including attempts to study union objectives empirically.

[2] Douglas Blair and David Crawford (1984) were perhaps the first to point out the relevance for studies of union behaviour of the general non-existence of voting equilibria when choices are multi-dimensional.

of the firm. Just as it is common to assume that shareholders are homogeneous in ways that matter for the firms' optimal behaviour, so modellers of unions almost always assume that union members differ along a single dimension at most. In addition, both union leaders and firm managers are commonly, although not always, assumed to be perfect agents of their constituents. In the theory of the firm, this leads to the assumption of profit maximization in the static case. In the theory of the union, this leads to the assumption that unions, in the static case, maximize a welfare function that depends on the wage of its members and the employment level in the sector the union covers. Denoting the union wage by w and the relevant employment level by L, unions are commonly assumed to maximize some variant of

$$u = u(w, L) \text{ with } \frac{\partial u}{\partial w} > 0 \text{ and } \frac{\partial u}{\partial L} \geqslant 0. \tag{1}$$

The unions' welfare is assumed to depend positively on the wage since a higher wage always benefits union members holding employment constant. Whether greater employment benefits union members holding the wage constant depends on whether or not union members are securely employed. If lay-offs are assumed to occur strictly by seniority and union members are assumed to vote according to their myopic self-interest, then the union would only care about the wage as long as more than half of the union membership remained employed. However, if the laid-off workers leave the union, union members with average seniority in the first period would have below-average seniority in the second (Farber 1986). Moreover, lay-offs rarely occur strictly according to seniority as union members who lose their job at one plant seldom have the right to take the job of a union member with less seniority at another plant. Thus, a majority of union members may feel threatened by lay-offs at union unemployment levels well below 50 per cent.

The debate over whether or not unions care about employment is intertwined with another debate over what union contracts cover. One convention, represented by Andrew Oswald (1982; 1985), among many others, is to assume that the labour agreement covers wages alone, with employment set by the firm in accordance with profit maximization. In this approach, the union maximizes $u(w, L)$ as given in equation (1) subject to the constraint that employment is given by the firms' demand for labour: $L = L(w)$ with $L'(w) < 0$. The other convention, represented by George De Menil (1971) and Ian McDonald and Robert Solow (1981), among others, assumes that both wages and employment levels are negotiated. This case can be represented by assuming that the union maximizes $u(w, L)$ subject to the constraint that the firms' profits, π, do not fall below some minimum value, or $\pi(w, L) \geqslant \pi_0$. In short, a contract that only covers the wage would produce outcomes along the demand-for-labour curve while a contract that covers both wages and employment might be expected to be located on the contract curve between the union and firms. Wasily Leontief (1946) was the first to point out that these two curves never coincide when the union cares about employment.

The most common argument in support of the assumption that contracts cover both wages and employment is that rational bargainers should seek to exploit all gains from trade. If the labour agreement does not cover both wages and employment, there exists another agreement that could make both union members and employers better off. The most common argument against this assumption is the observation that, in practice, union contracts rarely specify the employment level.

Neither argument is convincing. Against the claim that rational bargainers will not choose a point on the demand-for-labour curve, one can argue that (*a*) outcomes on the demand-for-labour curve are efficient from the bargainers' point of view if the union only cares about the wage (Oswald 1987), and (*b*) contracts off the demand-for-labour curve may not be incentive-compatible when firms have private information (Farber 1986). On the other hand, the fact that union contracts do not generally specify employment levels is not persuasive as union contracts do frequently cover work-rules that limit the firms' discretion to alter the capital–labour or the labour–output ratio. In such cases, the contract may force employers to choose higher levels of employment than they would like at the prevailing wage even though employment is not fixed explicitly (Hall & Lilien 1979). When major unions in the United States were asked to accept roll-backs in their contracts in the 1980s, the relaxation of work rules was high on the employers' list of demands.

One aspect of this question that has not been recognized in the literature is the relevance of the level at which bargaining occurs. Labour contracts that might implicitly cover employment by specifying work rules cannot be negotiated at the national level. Indeed, work rules must be negotiated at the plant level unless the industry is relatively homogeneous. Thus one way that decentralized bargaining can differ from centralized bargaining is in the scope of the labour agreement.

We will not pursue this possible difference, however. In order to illuminate other differences among bargaining levels, it helps to hold the coverage of union contracts constant. Thus we will assume throughout that firms choose the level of employment unilaterally, whether the contract is negotiated locally or nationally. Moreover, there is little of substance that is lost. The effect of bargaining over employment as well as wages can be illustrated in a model where firms alone set employment when there are hiring and firing costs, as we illustrate in Chapter 12.

Formally, then, the optimal union wage is modelled as the solution to

$$\max_{w} u(w, L(w)). \tag{2}$$

We will assume for most of Chapter 11 that $(\partial u/\partial L)$ is strictly positive and that the solution is given by the first-order condition

$$\frac{\partial u}{\partial w} + \frac{\partial u}{\partial L} L'(w) = 0. \tag{3}$$

What does this simple model tell us about wage demands in centralized versus decentralized bargaining systems? Consider the simplest possible case where all product prices are exogenous, i.e. given by the world market, so that w represents both the real and nominal wage. Assume, in addition, that there is a fixed number of firms in the economy, each with the same labour demand function $L(w)$. Thus $L(w)$ reflects the trade-off between wages and employment for the aggregate economy, as well as for any fixed subsector of the economy. Under these conditions, it is clear that the unions' optimal wage demand is independent of the degree of centralization.

(b) Endogenous Product Prices

The assumption that wages have no effect on product prices may be accurate for many industries in small open economies, but not for all. Where wage increases are passed on to prices to some extent, the unions' optimal wage is no longer independent of the level of centralization. This topic has been studied by Calmfors and Driffill (1988), Jon Strand (1989), and Michael Hoel (1991) in the context of a closed economy model in which both wages and prices are endogenous. The central result of this work is that the relationship between wages and bargaining level is hump-shaped, with both very decentralized and highly centralized bargaining systems producing greater wage restraint than bargaining systems in between.

To illustrate this result, we adopt the formulation of Hoel (1991). As before, we assume that there is a fixed number of firms with identical production functions. Product prices, however, are now assumed to be endogenously determined. Now we must distinguish between nominal and real wages. Let p be the product price of the industry under consideration. We assume that the price can be written in reduced form as a function of the nominal wage in the industry, w, and nominal wages elsewhere in the economy, denoted w^*:

$$p = p(w, w^*) \tag{4}$$

Similarly, product prices elsewhere in the economy, denoted p^*, are given by

$$p^* = p^*(w, w^*) \tag{5}$$

It is assumed that

$$\frac{\partial p}{\partial w}\frac{w}{p} \equiv \eta \in [0,1] \quad \text{and} \quad \frac{\partial p^*}{\partial w}\frac{w}{p^*} \equiv \eta^* \leqslant \eta.$$

An increase in w raises p, or leaves p unchanged as a special case, i.e. $\eta \geqslant 0$. It is also clear that a rise in w cannot have a larger impact in other sectors than in the sector with the wage increase, i.e. $\eta^* \leqslant \eta$. In the hypothetical case of a one-sector economy, we have $\eta = \eta^*$. More realistically we have $\eta^* < \eta$ when η is positive. It is not obvious whether a wage increase in one industry raises or lowers prices in other industries. If industries produce products that are complements in consumption, then an increase in w that produced an increase in p would reduce p^*. Thus, η^* could be either positive or negative.

Employment, as before, is a function of the real product wage $L = L(w/p)$. Workers' consumption possibilities, however, are a function of nominal wage divided by the consumer price index. The consumer price index, denoted p^c, is a function of prices throughout the economy:

$$p^c = p^c(p, p^*) \quad \text{with} \quad \frac{\partial p^c}{\partial p}\frac{p}{p^c} \equiv \theta \in [0, 1] \quad \text{and} \quad \frac{\partial p^c}{\partial p^*}\frac{p^*}{p^c} = 1 - \theta. \tag{6}$$

This last equation embodies the condition that a proportional increase of all prices increases the consumer price index by the same proportion.

We assume that wages are determined as the non-cooperative equilibrium of a wage-setting game among unions. That is, in equilibrium every union's wage is optimal given the wages chosen by other unions. Thus the problem facing each union is

$$\max_w u\left(\frac{w}{p^c(p(w, w^*), p^*(w, w^*))}, L\left(\frac{w}{p(w, w^*)}\right)\right). \tag{7}$$

In a symmetric equilibrium where $w = w^*$ and $p = p^* = p^c$, the first-order condition for (7) can be written as

$$\frac{\partial u}{\partial w/p^c}[1 - (\theta\eta + (1 - \theta)\eta^*)] + \frac{\partial u}{\partial L}L'(w/p)(1 - \eta) = 0 \tag{8}$$

or

$$\frac{\partial u}{\partial w/p^c} + h\frac{\partial u}{\partial L}L'(w/p) = 0 \tag{9}$$

where

$$h = \frac{1 - \eta}{1 - (\theta\eta + (1 - \theta)\eta^*)}. \tag{10}$$

The second-order conditions for a maximum imply that the optimal w is a negative function of h.

From (10), it can be seen that h is the ratio of two elasticities. The numerator of h is the elasticity of the real product wage with respect to the nominal wage chosen by the union. The denominator is the corresponding elasticity of the real consumption wage. It follows from $\eta^* \leqslant \eta$ that the denominator cannot be smaller than the numerator, which implies that $h \leqslant 1$. The important difference here between different degrees of centralization is the ability of each union to increase its real consumption wage without an equivalent increase in the real product wage in its sector. Since a rise in the real product wage reduces employment, the union wants this wage to increase as little as possible. If $h < 1$, each union can raise its real consumption wage proportionately more than the real product wage in its sector, resulting in a higher equilibrium real wage (in both senses) than it would have demanded had prices been exogenous.

Table 11.1 presents a comparison of how h, and therefore w and L, depends on the degree of centralization. Consider first the case of price-taking firms and wage-setting at the level of the firm. The wage in any single price-taking firm has a negligible effect on product prices, which implies that $\eta = \eta^* = \theta = 0$. In this case it follows from (10) that $h = 1$. In other words, the case of price-taking firms with decentralized wage-setting is identical to the case with exogenous prices. The interpretation is straightforward. It doesn't matter whether or not prices actually are exogenous. What matters is that each union perceives prices as independent of its own wage.

Consider next the opposite extreme, the case of perfectly

TABLE 11.1 *Wages and the Level of Wage-Setting*

Level of wage-setting	Price-taking firms	Monopolistic competition
Firm	$\theta = 0$	$\theta = 0$
	$\eta = \eta^* = 0$	$\eta > \eta^* = 0$
	$h = 1$	$h < 1$
Industry	$\theta \in (0, 1)$	$\theta \in (0, 1)$
	$\eta > \eta^*$	$\eta > \eta^*$
	$h < 1$	$h < 1$
Nation	$\theta = 1$	$\theta = 1$
	$h = 1$	$h = 1$

centralized wage-setting where $\theta = 1$. From $\theta = 1$, it follows immediately from (10) that $h = 1$ whatever the relationship between η and η^*. Thus the cases of complete decentralization and complete centralization give the same outcome. The reason is that in both cases each union bears the full consequences of a higher nominal wage itself. With wage-setting at the firm level, prices are perceived as fixed since firms are price-takers by assumption. With wage-setting above the firm level, the union chooses the nominal wage taking into consideration the effect of the nominal wage on prices. But since the industry real product wage and the real consumption wage are equal when bargaining is fully centralized, the outcome is unchanged.

With wage-setting at an intermediate level, the wage usually affects the product price of the sector the union belongs to, or $\eta > 0$. Since $\eta^* < \eta$ and, in this case, $\theta < 1$, we have $h < 1$. Wage-setting at an intermediate level thus produces higher wages and lower employment than wage-setting at either the firm or national level. The intuition behind this result is that each union knows that any increase in its nominal wage will increase its product price to a greater extent than it will raise the cost of living. This reduces the negative employment effects of an increase in the real consumption wage. Each union, in other words, is able to pass some of the cost of a wage increase on to others through the price effect, rather than bearing all of the cost itself in the form of lower employment. When all unions behave like this, however, the consequence is higher real product and real consumption wages and lower employment than would result from either highly decentralized or highly centralized wage-setting.

Consider now the case of imperfect competition where each firm faces a downward sloping demand curve. The qualitative result for the comparison of fully centralized wage-setting with wage-setting at an intermediate level remains the same as in the case of perfect competition. However, wage-setting at the level of the firm is no longer equivalent to wage-setting at the national level. With imperfect competition, a rise in the wage paid by a firm is to some extent passed on to prices, i.e. $\eta > 0$. In this case, firm level wage-setting produces higher real wages than nation-wide wage-setting, a result first derived by Cahuc (1987). The interpretation is similar to the case of industry-level wage-setting. With monopolistic competition, the firm-level union knows that if it raises its wage, the wage increase will to some extent be passed on to the price of the firm's product. This reduces the negative employment effect of the wage increase. As in the case of intermediate-level bargaining, the end result is higher wages and lower employment than unions would have chosen if they chose wages jointly.

To summarize the results so far, with exogenous prices, complete centralization or firm-level wage-setting with price-taking firms, a union that raises its nominal wage affects its sector's real product wage and its real consumption wage in the same proportion. With intermediate-level wage-setting or with fully decentralized wage-setting under conditions of imperfect competition, a nominal wage increase raises the real product wage of its sector proportionately less than its real consumption wage. Since the benefits of higher wages come from the real consumption wage while the costs of higher wages come from the real product wage, the gap between the two induces the union to choose higher nominal wages. When all unions do the same, both prices and real wages are higher and employment is less than unions would choose had they been able to co-ordinate their wage demands.

(c) Substitutes and Complements in Production

A second way in which the simple model of section (a) is unrealistic is in the assumption that each product is made with the labour of a single union. Final products, in general, depend on many different types of labour that are often represented by different unions. Firms frequently bargain directly with more than one union. This is particularly true in industries and countries where blue-collar workers are organized in craft unions or in competing industrial unions. In large metalworking firms in Britain, for example, it is not unusual for the work-force to be represented by 15–20 unions (Bratt 1986). Even in countries like Norway and Sweden where non-competing industrial unions are the rule, there are separate unions for blue-collar, white-collar, and professional workers.

In addition, firms depend on the labour of workers they do not directly employ. Payments for goods and services bought from other domestic producers may comprise a substantial part of a firm's production costs. The manufacturing sector depends on the outputs of workers in utilities and transportation. The cost of new investment depends on the price of capital goods and new construction. The cost of government services depends on wages in the public sector. According to the comment by Stephen Nickell (Calmfors & Driffill 1988: 52), labour costs average only 20 per cent of revenues at the firm level in Great Britain yet wages and salaries constitute 70 per cent of value added at the national level.

When products are produced by workers divided into multiple unions, the unions' optimal wage depends on the level of centralization even when final product prices are fixed in world markets. Wage-setting by multiple types of workers organized in separate unions was first studied by Sherwin Rosen (1970), but the topic received relatively little attention until recently. Oswald (1979) examined the existence of equilibria in an economy with multiple unions. Henrik Horn and Asher Wolinsky (1988) and Tor Hersoug (1985) studied the question of the optimal division of workers into separate unions (from the workers' point of view) and highlighted the critical distinction between complements and substitutes in production.

Matti Pohjola (1984) and Michael Wallerstein (1990) studied the impact of decentralized versus centralized bargaining with different types of labour within a differential game framework. Here we illustrate their basic results using the simpler static framework adopted by Oswald (1979).

Let there be k unions whose labour is used in production. The product price is assumed to be exogenous (i.e. determined by the world market). The interdependence of the k unions can be represented in reduced form by letting the demand for labour for each union be a function of all k wages: $L_i = L_i(w_1, \ldots, w_k)$.[3] With decentralized wage-setting, union 1's problem is

[3] If the profit of the firm is written as $\pi = pF(L_1, \ldots, L_k) - \Sigma w_i L_i$ where $F(L_1, \ldots, L_k)$ is the production function with k types of labour, the first-order condition for employment of members of the ith union is $(\partial F/\partial L_i) = (w_i/p)$ which gives $L_i = L_i(w_i, L_1, \ldots, L_{i-1}, L_{i+1}, \ldots, L_k)$. Doing the same for all k unions, one can use the k equations to eliminate the variables $(L_1, \ldots, L_k)$ from the right-hand-side and write $L_i = L_i(w_1, \ldots, w_k)$ as in the text.

$$\max_{w_1} u_1(w_1, L_1(w_1, \ldots, w_k)) \tag{11}$$

with the first-order condition

$$\frac{\partial u_1}{\partial w_1} + \frac{\partial u_1}{\partial L_1}\frac{\partial L_1}{\partial w_1} = 0. \tag{12}$$

The second-order condition for a maximum implies that the left-hand-side of (12) is a negative function of w_1.

If, in contrast, the k wages were chosen jointly to maximize some collective welfare function V that depends positively on the welfare of each of the k unions, the collective choice is given by the solution to the problem

$$\max_{w_1, \ldots, w_k} V(u_1(w_1, L_1(w_1, \ldots, w_k)), \ldots, u_k(w_k, L_k(w_1, \ldots, w_k))). \tag{13}$$

The first order condition for (13) that corresponds to (12) is

$$\frac{\partial V}{\partial u_1}\left[\frac{\partial u_1}{\partial w_1} + \frac{\partial u_1}{\partial L_1}\frac{\partial L_1}{\partial w_1}\right] + \sum_{i=2}^{k}\frac{\partial V}{\partial u_i}\frac{\partial u_i}{\partial L_i}\frac{\partial L_i}{\partial w_1} = 0 \tag{14}$$

The terms $(\partial V/\partial u_i) > 0$ represent the weight given to each union in the aggregate welfare function. As long as unions care about employment of their own members or $(\partial u_i/\partial L_i) > 0$, it can be seen from (14) that the effect of centralization depends on the impact of the first union's wages on the demand for the labour of members of other unions: $(\partial L_i/\partial w_1)$

While $(\partial L_i/\partial w_i)$ is negative for all i, the terms $(\partial L_j/\partial w_i)$ for $i \neq j$ may have either sign. If $(\partial L_j/\partial w_i) > 0$, the two unions are said to be substitutes in production. A higher wage for union i induces the firm to employ more members of union j. On the other hand, if $(\partial L_j/\partial w_i) < 0$, the two unions are said to be complements. A higher wage for one reduces the firm's demand for the labour of the other. In this case, the two unions are supplying complementary labour in the sense that the productivity of each is enhanced for the presence of the other. Horn and Wolinsky (1988) argued that workers who are substitutes have an incentive to organize into a single union since, by uniting, they increase their ability to strike effectively. In contrast, workers who are complements increase their joint bargaining power by remaining in separate unions, since the cost to the firm of separate strikes exceeds the cost of a strike by the two groups simultaneously. If the division of workers into unions reflects choices made to maximize bargaining power, members of separate unions would be complements rather than substitutes in production.

To simplify the exposition, consider the case with identical and symmetrical unions such that $\partial L_i/\partial w_i = \partial L/\partial w$ and $\partial L_i/\partial w_j = \partial L/\partial w^*$ for all $i, j = 1, \ldots, k$ with $i \neq j$. Suppose, in addition, that a centralized wage-setting chooses wages to maximize the average welfare of the k unions, or equivalently when k is fixed, the sum of the unions' welfare: $V = \Sigma u_i$. In this case, the optimal wage for the k unions bargaining jointly is given by

$$\frac{\partial u}{\partial w} + \frac{\partial u}{\partial L}\left[\frac{\partial L}{\partial w} + (k-1)\frac{\partial L}{\partial w^*}\right] = 0 \tag{15}$$

In contrast, the non-cooperative equilibrium of decentralized wage-setting is given by equation (12) without subscripts:

$$\frac{\partial u}{\partial w} + \frac{\partial u}{\partial L}\frac{\partial L}{\partial w} = 0 \tag{16}$$

Since the left-hand-side of (16) is a negative function of the wage, the optimal wage with centralized wage-setting is lower than the non-cooperative equilibrium of decentralized wage-setting if members of different unions are complements (i.e. if $(\partial L/\partial w^*) < 0$). If workers are substitutes (i.e. if $(\partial L/\partial w^*) > 0$), centralization would raise the unions' wage demands.

When one union's wage affects other unions' wage and employment possibilities, it is apparent that decentralized wage-setting differs from centralized wage-setting. When different unions are substitutes in production, each union's wage increase raises the demand for the labour supplied by other unions. A centralized wage-setter that internalized this benefit would want to raise wages above the decentralized equilibrium. More commonly, one union's wage increase reduces the demand for the labour supplied by other unions. In this case, centralized wage-setting would reduce wage demands below the equilibrium wage demanded by unions acting independently.

This result appears to strengthen the case for highly centralized bargaining and weaken the case for very decentralized bargaining. A better interpretation, in our opinion, is that it is incorrect to think that bargaining systems can be arrayed along a single dimension of centralization. Union members can be divided into separate organizations in a variety of ways with differing consequences for the effects of decentralized wage-setting. A decentralized bargaining system comprised of company unions, as in Japan, is quite different from a decentralized bargaining system comprised of multiple craft and competing industrial unions, as in the United States or Great Britain.

There are at least two dimensions of centralization that ought to be distinguished in empirical work but never are. The first dimension is whether wages are set at the level of the plant, enterprise, industry, or nation. The second dimension is whether workers in different types of jobs bargain jointly or separately. Putting the two together, one obtains something like Table 11.2. As one moves vertically down the table, the relationship between wage demands and centralization is likely to be hump-shaped according to both the model with endogenous final prices and the model with different types of labour. In terms of the endogenous price model, industry-level wage-setting maximizes the extent to which the cost of a wage increase can be passed on to others as a price increase. In terms of the model with different types of labour, workers doing similar jobs at different plants or enterprises in the same industry are typically substitutes while workers in different industries are more likely to be complements. But as one moves horizontally across the table, the relationship between the militancy of wage demands and centralization is mono-

TABLE 11.2 *Dimensions of Centralization*

Levels of wage-setting	Each type bargains separately	All types bargain jointly
Plant	Complete decentralization	Company unions
Enterprise		
Industry	Craft unions	Industrial unions
Nation		Complete centralization

tonically declining as workers in different types of jobs are most often complements.

(d) Other Externalities in Wage-Setting

There are a number of different possible externalities in wage-setting that can be captured by writing the unions' maximand as

$$u = u(w, L(w), z(w, w^*)) \tag{17}$$

where z is some variable that depends on wages elsewhere in the economy, w^*, as well as wages in the plant. For example, observers of industrial relations have long claimed that workers care about wage differentials in addition to wage levels. Economists often go to great lengths to avoid adding a concern with relative income as an argument in workers' utility functions, for both good and bad reasons.[4] The more freedom one has to make *ad hoc* adjustments to workers' utility, the easier it is to demonstrate any conclusion one wants. At the same time, the simplifying assumption that workers care only about their own income (and leisure) that is made for analytic convenience should not be mistaken for reality. Workers may strive for status as well as income and status may depend on relative income. Or workers may be concerned with notions of fairness that are derived from comparisons with what others are paid (Elster 1989).

Suppose, for whatever reason, that union members care about how much they are paid relative to other workers in addition to the standard concerns with wage levels and employment security. Then z in (17) could be written as $z = w/w^*$ with $(\partial u/\partial z) > 0$. If all unions try to increase their wage relative to the wages of others, none will change position (provided their relative bargaining strength has not changed). Wages will increase, however, and unemployment will rise. According to this reasoning, centralized wage-setting reduces wages by inhibiting the fruitless struggle of each group to raise its wage more than the others.

[4] Not all economists have ignored workers' concern with relative wages. See Oswald 1979; Gylfason & Lindbeck 1984; and Uddén-Jondal 1989; 1990, for formal analyses of the consequences of envy on wage-setting.

A different interpretation of z centres on aggregate unemployment. Union members care about the aggregate rate of unemployment to the extent that they face some positive probability of losing their job. Every additional job-searcher reduces the likelihood that other job-seekers will find work (Mortensen 1986). Even union members who considered their jobs secure would care about aggregate unemployment to the extent that government expenditures to support the unemployed result in higher taxes on employed workers (Jackman 1990; Holden & Raaum 1989). In either case we might interpret z as the economy-wide rate of unemployment with $(\partial u/\partial z)<0$. With local wage-setting, $(\partial z/\partial w)$ is close to zero. As the coverage of the labour agreement expands, $(\partial z/\partial w)$ increases implying a lower optimal wage.

A third interpretation of z is the probability of having a social democratic government. Union members would care about the party in power to the extent that social democratic governments are more likely than bourgeois governments to adopt policies that favour union members. Union leaders may care about the party in power because they have close personal links with the social democratic party leadership. According to this argument, it is the government, not the unions, that takes responsibility for aggregate unemployment. The unions, however, care about the survival of the government if it is social democratic. In this case, centralized wage-setting reduces wage demands relative to decentralized wage-setting under social democratic governments but not under bourgeois governments, as argued by Lange and Garrett (1985) and Garrett and Lange (1986).

All of these externalities have a similar implication: with centralized bargaining, wage-setters internalize the impact of the wage agreement on relative wages and/or aggregate unemployment which leads them to moderate their wage demands. Of course, it is unrealistic to think that a centralized union confederation has the capacity accurately to assess all of the externalities in wage-setting and choose the optimal national wage schedule. Yet, if all externalities point in the same direction, centralized union negotiators may indeed be satisfied with lower wage levels than would be the outcome of decentralized wage-bargaining.

(e) Change of Preferences

So far, we have not yet systematically considered the possibility that local unions and centralized confederations might have different objectives. Yet recent work that emphasizes the importance of insiders versus outsiders in the theory of union behaviour points to an important difference between decentralized and centralized bargaining that stems from the way centralization affects the definition of insiders.

To examine this topic we drop the assumption that $(\partial u/\partial L)$ is strictly positive for all levels of L. It is more reasonable to assume, as in the insider–outsider model, that the willingness of a union to accept a lower wage for greater employment depends on the extent to which current union members are unemployed. In an expanding industry where the demand for labour exceeds the current union membership, union members have little reason, apart from altruism, to restrain their wage demands to enable employment to grow even faster. On the other hand, in a declining industry where union members face lay-offs, the union may well sacrifice wage gains to reduce or halt the employment decline. The effect of the size of the union relative to the demand for labour on union behaviour can be represented by writing the union's objective function as

$$u = u(w, \max(M - L, 0)) \tag{18}$$

where M is the union's current membership and L is the demand for labour (Oswald 1987; Wallerstein 1987). Equation (18) implies that the union's indifference curves are downward sloping where $L < M$ but flat for $L \geqslant M$ as illustrated by the curves ABC and $A'DE$ in Figure 11.1. If the demand-for-labour curve is represented by the line LL, the optimal wage is w and union members are fully employed.

The line between insiders and outsiders may depend on the level of bargaining. For the local union, the distinction seems clear. Insiders are current members of the local union. If, after some time, unemployed union members leave the union and new employees do not join the union immediately (or are not considered full members by the union at first), then the union's current membership is given by past employment in the plant.

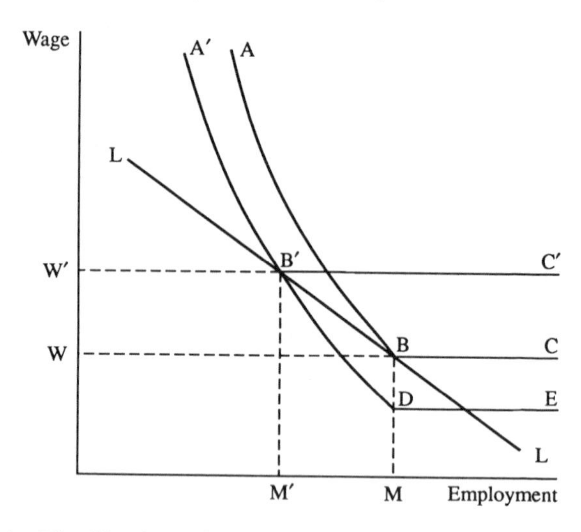

FIG. 11.1 The Number of Insiders and Union Preferences of Wages and Employment

Let the time it takes unemployed workers to quit the union be equal to the time it takes new employees to become union members. Then L and M in the case of decentralized bargaining are given by

$$L = L^i \quad \text{and} \quad M = L^i_{-t} \tag{19}$$

where L^i is employment in plant i and t is the amount of time it takes for the unemployed to quit and the newly employed to join.

An important implication of equation (19) is that there is hysteresis in unemployment: the current equilibrium unemployment level depends on the past level of unemployment (Gottfries & Horn 1986; Blanchard & Summers 1987; Lindbeck & Snower 1988). An unforeseen decline in demand that causes lay-offs lasting for more than t periods reduces the union membership and thus reduces the threshold employment level above which the union only cares about wages. An unforeseen increase in demand for more than t periods has the opposite effect of increasing union membership and raising the union's sensitivity to unemployment.[5]

[5] The symmetry of increases and declines in demand is a consequence of the assumption that the time it takes new workers to become union members equals

The distinction between insiders and outsiders is less clear at the national level. According to Tor Hersoug, Knut Kjær, and Asbjørn Rødseth (1986), the central Norwegian trade union confederation (LO) has no statistics on unemployed members and no way of deriving such statistics from official sources. Thus it is the national unemployment rate that enters in the LO's calculations of the employment consequences of its wage demands. More generally, we propose the following general formulation for national wage-setting:

$$L = \sum_i L^i$$

$$M = \sum_i L^i_{-t} + \gamma_1\left(N_{-t} - \sum_i L^i_{-t}\right) + \gamma_2(N - N_{-t}) \quad \text{with} \quad 1 \geqslant \gamma_1 \geqslant \gamma_2 \geqslant 0 \qquad (20)$$

The first term of the second line, ΣL^i_{-t}, is the current union membership. The second term, $N_{-t} - \Sigma L^i_{-t}$, consists of older workers who have not been employed for t periods. The third term, $N - N_{-t}$, consists of new workers who have just entered the work-force. The measure M is based on the idea that the central union does not only care about whether its core members are employed or not. Some weight is also placed on the job opportunities of unemployed older members and new workers who have not yet become members of the union.

Even if the central union did not care at all about the second two groups of workers, i.e. $\gamma_1 = \gamma_2 = 0$, centralized wage-setting might differ from decentralized wage-setting. Unemployment is always distributed unevenly. Some industries continue to grow while others decline. On the one hand, national bargainers would be more sensitive to unemployment confined to a few industries than would local negotiators elsewhere in the economy. On the other hand, national bargainers may be less sensitive than would local negotiators in the declining industries. Which effect would be stronger is not clear.

In practice, $1 = \gamma_1 > \gamma_2 > 0$ seems to be a more accurate characterization of the preferences of central wage-setters in the Nordic countries. The national confederations care equally about

the time it takes laid-off workers to leave the union. See Lindbeck & Snower 1988 for a discussion of symmetrical and asymmetrical hysteresis.

unemployment of members and non-members, $\gamma_1 = 1$, if only because it may be difficult for the central confederation to distinguish the two groups as Hersoug, Kjær, and Rødseth argue. The central confederations also seem to care to a lesser extent about new entrants in the labour market, $\gamma_2 > 0$, perhaps because of the political ties between the leadership of the blue-collar union confederation and the social democratic party. Thus, for a variety of reasons, centralized bargainers appear to have a broader definition of insiders than local bargainers. A broader definition of insiders, in turn, leads directly to a greater willingness to reduce wage demands for greater employment.

This is illustrated in Figure 11.1. Let the number of insiders be M' for local bargaining, but M for centralized bargaining, perhaps because there are lay-offs in other plants. Indifference curves of the central bargainers are given by ABC or $A'DE$ as before. The indifference curve for the local union is $A'B'C'$, however. With LL as the demand for labour, the central bargainers would choose the wage of w while local bargainers would set the wage at w'. The fact that centralized bargainers have a broader constituency than decentralized bargainers increases the sensitivity of centralized wage-setting to unemployment.

(f) Efficiency Wage Considerations

Throughout the discussion, our attention has been focused exclusively on the unions' wage demands. The implicit assumption in most of the literature is that employers only benefit from centralized wage-setting to the extent that centralization moderates union wage demands. If unions lost their influence over wages, it is usually thought that all rationales for centralized wage-setting disappear. Recent work, however, on decentralized and centralized wage-setting incorporating the effect of wages on productivity by Hoel (1989*a*) and Rødseth (1990) suggests that the centralization of wage-setting might reduce wage levels and increase employment even if wages were unilaterally set by employers instead of unions.

The basic premiss of a wide class of efficiency wage models is that workers' efficiency, denoted e, is a positive function of their wage relative to wages and employment possibilities elsewhere.

If, for any reason, the efficiency of labour is affected by the wage, then employers may find it optimal to pay wages higher than the market-clearing level. Let the firm choose both employment and the wage rate. Then the firm's decision can be written as

$$\max_{w,L} R(e(w)L) - wL \tag{21}$$

where $R(\cdot)$ is the firm's revenues. In the case of an interior solution, the first order conditions are

$$R'(eL) - w/e = 0, \tag{22}$$

$$R'(eL)\frac{de}{dw} - 1 = 0. \tag{23}$$

Let $\lambda(w) \equiv (de/dw)(w/e)$ be the elasticity of workers' efficiency with respect to the wage. Then equations (22) and (23) can be combined to give

$$\lambda(w) - 1 = 0 \tag{24}$$

as the basic optimality condition in the efficiency wage model. The second order condition for a maximum implies that $(d\lambda/dw) < 0$.

There are many possible reasons why e might depend on w.[6] For example, a higher relative wage might lower turnover and thus reduce the costs associated with finding and training new workers (Calvo 1979). Or a higher relative wage increases the loss associated with being fired and thus may reduce shirking on the job (Calvo & Wellisz 1978; Shapiro & Stiglitz 1984; Bowles 1985). In either case, what matters is the difference or, more conveniently, the ratio between a worker's current wage w and what a worker would obtain if he or she quit or were fired. We assume for convenience that the probability of finding another job after a separation is equal to one minus the aggregate rate of unemployment. Workers' expected income after a separation can then be written as $\mu w^* + (1 - \mu)bw^*$ where w^* is, as before, the wage level elsewhere, b is the replacement ratio (the percentage of wage income that is replaced by unemployment benefits), and μ is the employment rate (one minus the rate of unemployment).

[6] See Akerlof & Yellen 1986 for a collection of papers that describe the relationship between wages and productivity in many different ways.

Thus, we have

$$e = e\left(\frac{w}{w^*[\mu + (1-\mu)b]}\right) \quad \text{with} \quad e'(\cdot) > 0 \tag{25}$$

as the equation representing the dependence of efficiency on wages.

Workers' outside opportunities, the denominator in equation (25), is exogenous from the point of view of each employer. Therefore, the elasticity of efficiency with respect to the wage in the case of decentralized wage-setting is

$$\lambda^D = \frac{we'/e}{w^*[\mu + (1-\mu)b]} \tag{26}$$

which firms set equal to one by (24). As each employer tries to raise wages relative to others, none succeeds but the aggregate wage level and rate of unemployment increase until (24) is satisfied.

With centralized wage-setting, all wages are raised together. When $w = w^*$, both terms drop out of the expression for e in (25). At the same time, centralized employers would take into consideration the effect of higher wages on unemployment: $\mu = \mu(w)$ with $\mu'(w) < 0$. As Michał Kalecki (1943) argued, employers benefit from higher unemployment to the extent that it increases the 'threat of the sack'. Calculating the elasticity of workers' efficiency with respect to centrally set wages evaluated at the optimal wage in the decentralized equilibrium, one obtains

$$\lambda^C = \lambda^D \left[\frac{\mu(1-b)}{\mu(1-b)+b}\right]\left(\frac{-w\mu'}{\mu}\right). \tag{27}$$

This last equation can be simplified further, assuming a fixed number of identical firms. Let ξ be the elasticity of the demand for efficiency units of labour.[7] Then

$$\left(\frac{w\mu'}{\mu}\right) = \xi(1-\lambda) - \lambda = -1 \tag{28}$$

since, in the decentralized equilibrium, $\lambda = 1$ by equation (24). Therefore, we have

[7] Let $G(w/e) \equiv R'^{-1}$ be the demand for efficiency units of labour eL. Then $\xi \equiv (wG'/eG)$.

$$\lambda^C = \lambda^D \left[\frac{\mu(1-b)}{\mu(1-b)+b} \right] < \lambda^D. \tag{29}$$

Thus $\lambda^C - 1 < 0$ when evaluated at the decentralized equilibrium wage. To achieve the optimum of $\lambda^C = 1$ (provided there is only one wage level that satisfies this condition), the wage must be reduced since $(d\lambda/dw) < 0$ by the second-order condition. A centralized confederation of employers would therefore set lower wages through a national agreement than would be chosen by each employer separately. Total profits increase with centralization: the direct gain to employers of avoiding the attempt by each to raise wages above wages elsewhere outweighs the indirect loss of decreased discipline due to lower unemployment. To the extent that unions also care about efficiency or turnover (short-term workers are harder to recruit in the union), the same model could be applied when the union sets the wage as well.

12

Bargaining Models of Wage-Setting

(a) Models of Bargaining over Wages

In reality, unions rarely set wages unilaterally. Neither do firms in the unionized sector of the economy. The labour contract is the result of a bargaining process in which the two sides must reach an agreement. To write, as we sometimes did, of the resulting wage was a convenient shorthand for the wage that would result if firms were forced to accept the unions' demands (or if unions were forced to accept the firms' demands). Thus the models developed in the previous chapter should be viewed as studies of bargaining goals rather than as studies of the actual outcome of bargaining. These models are useful as models of the effect of the bargaining structure on the unions' willingness to exercise self-restraint and on the firms' desire to hold down wages.

Nevertheless, many aspects of wage-setting cannot be understood without a model of the bargaining process. The basic problem in collective bargaining is how the quasi-rents that are inherent in the employment relationship should be divided between workers and employers. The first question to be studied is the relationship, if any, between the level of centralization and the division of the surplus. An important related question is the extent to which mixed systems where wage increments or drift are negotiated locally after a base wage is set centrally are really centralized at all.

The question of how the structure of bargaining affects the surplus to be bargained over is as important as the question of how bargaining structure affects the way the surplus is divided. This leads to two separate issues of efficiency. The first is the efficiency of the bargaining process itself or the ability of bargainers to reach agreement without engaging in strikes or lock-outs. The second issue, perhaps the most important, concerns the impact of the centralization of bargaining on decisions regarding other variables that are not bargained over.

In the previous chapter, we adopted an exceedingly simple assumption about bargaining—that the wage is set unilaterally—in order to examine in greater detail the wage level unions or firms would prefer. In this chapter we simplify the assumption regarding the unions' objectives in order to study the effects of the bargaining process in different bargaining systems. We assume throughout this section that unions seek to maximize the wage (or the wage minus the disutility of labour) received by employed union members. Therefore, the models that follow are models of how the level of bargaining affects economic performance independently of its effects on the unions' willingness to accept wage restraint or on the firms' optimal wage policy.

The bargaining problem has been fruitfully studied in both co-operative and non-cooperative game theory. The first approach, inaugurated by John Nash (1950), was to consider bilateral bargaining as a co-operative game. The distinctive assumption of co-operative games, that binding agreements are feasible, seems appropriate in the context of bargaining over a legally enforceable labour contract. Nash defined the bargaining problem as consisting of a set of feasible agreements and a pair of disagreement pay-offs specifying what each side would obtain in the absence of an agreement. The problem is to determine what agreement will be reached.

Nash's method was axiomatic. He presented a list of axioms that a reasonable solution should satisfy and then proved that the axioms uniquely determined a particular solution. Nash assumed that the solution must be both individually and collectively rational. That is to say, the agreement must be no worse for each than no agreement and Pareto optimal. A third axiom, highly questionable but common to most of co-operative bargaining theory, is that the agreement should not depend on interpersonal comparisons of utility, although the solution depends on each side's attitude towards risk (Roth 1979). Nash added a fourth axiom, called independence of irrelevant alternatives, that states that if two games share the same pair of disagreement pay-offs, if the feasible set of one game is contained inside the feasible set of the other, and if the solution of the bigger game is attainable in the smaller game, then both games must have the same solution. These axioms are sufficient to determine a unique solution with a very simple mathematical structure.

Let R represent the revenue of the firm. Temporarily ignoring fixed costs, we can write the profits of the firm as $\pi = R - wL$ where w is the union wage. Let the disagreement pay-offs be written $\tilde{u}$ for the union and $\tilde{\pi}$ for the firm. Then the generalized Nash bargaining solution is found by solving the problem

$$\max_{w}(w - \tilde{u})^{\alpha}(R - wL - \tilde{\pi})^{1-\alpha} \quad \text{with} \quad \alpha \in [0, 1] \tag{1}$$

which yields

$$w = \alpha\left[\frac{R - \tilde{\pi}}{L}\right] + (1 - \alpha)\tilde{u} \tag{2}$$

as the wage. Thus the solution has the reasonable form of a weighted average between the best the union could hope to obtain, i.e. a wage such that $\pi = \tilde{\pi}$, and the worst, $\tilde{u}$. Nash originally added another axiom of symmetry that fixed $\alpha = 1/2$. It is more common in the literature to interpret α as a measure of bargaining power that can take any value from zero to one. Many other co-operative solutions to the two-person bargaining problem based on different sets of axioms have since been proposed, but Nash's solution remains the most commonly used.[1] Moreover, almost all solutions produce the same wage equation (2) for this simple bargaining problem.

According to co-operative game theory, the outcome is determined by the set of feasible agreements, R, the disagreement pay-offs, $\tilde{u}$ and $\tilde{\pi}$, and the measure of bargaining power, α. Thus we would like to know how each is affected by the level of centralization. However, co-operative bargaining solutions either assume $\alpha = 1/2$ or treat α as exogenous. Neither is satisfactory. In addition, there is an ambiguity in the definition of $\tilde{u}$ and $\tilde{\pi}$ that becomes apparent once the model is applied. In the absence of an agreement, are workers on strike or are they working at other, possibly non-union, jobs? Should $\tilde{u}$ be set equal to strike support or the competitive wage?

[1] See Roth 1979 and Kalai 1985 for surveys of the different co-operative solutions to the two-person bargaining problem.

These gaps in co-operative bargaining theory have, to some extent, been filled by the newer non-cooperative approach developed by Ingolf Ståhl (1972) and Ariel Rubinstein (1982). The essence of the non-cooperative approach is to write an explicit representation of the negotiation process as an extensive-form game and look for the equilibrium. The advantage of such an approach is that the solution is derived from optimizing behaviour rather than from a set of axioms that may or may not appeal to the reader's intuitions. Moreover, the non-cooperative approach, by forcing the modeller to be explicit about who can do what and when, opens up the study of the effect of such things as the sequence of moves on the outcome. On the other hand, the disadvantage of the non-cooperative approach is that the outcome is generally sensitive to small changes in arbitrary assumptions regarding the minutiae of the bargaining process. The price of greater explicitness is less generality.

Nevertheless, recent advances in non-cooperative bargaining theory have been valuable and widely adopted in models of wage negotiations. Rubinstein modelled the bargaining process as an extensive-form game in which the opposing sides make alternating offers.[2] Each side is restricted to making one offer every other period. After an offer is made, the opponent can either accept or reject. If the offer is accepted the game ends and the agreement is implemented immediately. If the offer is rejected, then the one rejecting the offer makes the next offer after a delay of one period. The game continues in this way until an offer is accepted. Rubinstein assumed, critically, that waiting is costly. In the bargaining model we consider here, both sides are assumed to discount future pay-offs over an infinite horizon.

The principle of subgame perfection states that players cannot bind themselves to take future actions that they would prefer not to take once the future arrives. Put another way, subgame perfect equilibria are equilibria supported by credible threats. It is not clear that the restriction of subgame perfection is always reasonable. Sometimes actors do seem to commit themselves to follow through on threats that would injure themselves as well as their opponent in order to obtain an advantage in bargaining. However, Rubinstein demonstrated that the restriction to

[2] See Sutton 1986 for a good introduction to the Rubinstein bargaining model.

subgame perfect equilibria of his extensive-form game resolved the conundrum-blocking development of a non-cooperative approach to bargaining: it produced a unique solution.

Avner Shaked and John Sutton (1984) provided an intuitive description of the solution by starting with each side's optimal strategy. Suppose it is the union's turn to make an offer. What should the union demand? One plausible answer is that the union should ask for as much as possible without asking for so much that the firm would gain by turning the offer down. Write the lowest wage the firm can hope to obtain in the next round, when it will be the firm's turn to make the offer, as w_F. Then the highest wage the firm would not reject in the present round, w_U, is the wage that leaves the firm indifferent between accepting or rejecting and obtaining w_F one period later. If the discount factor used by the firm is $\delta_F \in (0, 1)$, then, with an infinite horizon, the highest union offer the firm will accept is given by

$$\frac{1}{1-\delta_F}(R - w_U L) = \tilde{\pi} + \frac{\delta_F}{1-\delta_F}(R - w_F L). \tag{3}$$

The left-hand side of (3) is the current value of the firm's profits at the wage w_U. The right-hand-side is the similar sum when the firm must endure a disagreement for one period and then obtain the profits associated with the wage w_F.

But how should w_F be determined? When it is the firm's turn to make the offer, similar reasoning suggests that the firm will offer to pay as low a wage as possible without offering a wage so low that the union is better off rejecting the offer. If the best the union can obtain when it makes the offer is w_U, the best the firm can do is to offer the wage given by

$$\frac{1}{1-\delta_U} w_F = \tilde{u} + \frac{\delta_U}{1-\delta_U} w_U \tag{4}$$

where $\delta_U \in (0, 1)$ is the discount rate of the union. Since the game is identical every time the union makes an offer, equations (3) and (4) can be combined to yield

$$w_U = \left[\frac{1-\delta_F}{1-\delta_U\delta_F}\right]\left(\frac{R-\tilde{\pi}}{L}\right) + \left[\frac{\delta_F(1-\delta_U)}{1-\delta_U\delta_F}\right]\tilde{u} \tag{5}$$

as the solution when the union makes the first offer. The union cannot do better than offer w_U in the first period and the firm cannot do better than accept the offer.

According to (5), the union gets an arbitrary advantage by being designated as the first mover, since the firm must pay the cost of disagreeing for one period before it can make a counter-offer. This first-mover advantage disappears, however, as the time interval between offers goes to zero. Rewriting the discount factors as $\delta_F^t = (1 + (\rho_F/\Delta))^{-\Delta t}$ and $\delta_U^t = (1 + (\rho_U/\Delta))^{-\Delta t}$ where Δ is the time between offers and taking the limit as $\Delta \rightarrow 0$, one obtains $w_U = w_F = w$ with w given by

$$w = \left(\frac{\rho_F}{\rho_F + \rho_U}\right)\left(\frac{R - \tilde{\pi}}{L}\right) + \left(\frac{\rho_U}{\rho_F + \rho_U}\right)\tilde{u}. \tag{6}$$

Comparing (6) with (2), it can be seen that the non-cooperative approach provides an interpretation of the measure of bargaining strength α. According to (6), bargaining strength is a function of the relative impatience of the two sides to reach an agreement. The less impatient the union is relative to the firm, the larger the share of the pie the union receives and vice versa. There does not seem to be a general reason why the impatience of either the unions or firms should vary with the level of centralization. This implies that centralization does not affect α. If centralization affects the distribution of the firm's revenues between profits and wages, it must be because centralization alters the disagreement pay-offs.

The non-cooperative model also has strong implications regarding the interpretation of $\tilde{\pi}$ and $\tilde{u}$. It is clear from (3) and (4) that $\tilde{\pi}$ and $\tilde{u}$ should be interpreted as the income received by the firm and the union during a conflict.[3] The 'outside options', the wage workers could get at other jobs and the profits the firm could obtain by hiring new workers, has no impact on the

[3] More precisely, $\tilde{\pi}$ and $\tilde{u}$ are the income that would be received by the firm and the union during the bargaining process before an agreement is reached. Recent work has raised the question of what happens if the game is expanded to include the choice of strike or continue to work whenever an offer is rejected. The troubling answer is 'almost anything'. See Haller & Holden 1990 and Fernandez & Glazer 1991.

solution other than as constraints (Sutton 1986).[4] In order to attract workers, firms must pay at least as much as their workers can obtain elsewhere. Similarly, in order to maintain their jobs, the union cannot reduce profits so far that the firm would be better off shutting down the plant. In local bargaining, the outside options may be binding for less productive firms. With more centralized bargaining, the constraints are unlikely to bind, and the outside options are unlikely to influence the outcome of wage negotiations.

Thus centralization primarily affects the sharing of the pie in so far as centralization alters the conflict pay-offs of the two sides. Our earlier discussion of complements and substitutes is relevant here. The uniting of substitutes increases the share the union can obtain by reducing $\tilde{\pi}$. When groups are substitutes, the cost to the firm of a joint strike is more than the sum of the costs of separate strikes by each. When groups are complements, the cost of a joint strike is less than separate strikes. Unions usually prefer industry unions to separate company unions since workers in different firms in the same industry are substitutes in production.

Once an industry-wide union exists, employers often prefer industry-wide bargaining. Strikes against one firm at a time are more costly to the firm (since it loses business to its competitors) and less costly to the union (since the local can obtain strike support from the rest of the union) than would be a strike against all at once. Thus industry-level bargaining has frequently been sought by employers in order to reduce the union's bargaining power.

(b) Two-Tiered Bargaining

It is impossible to have centralized wage-bargaining without supplementary local bargaining. Some issues, like working conditions inside the plant, are inherently local. Even wage scales need to be adjusted according to local needs. Thus all centralized bargaining systems depend on supplementary local bargaining, if only over the implementation of the central agreement. The

[4] Erling Barth (1991) has shown that the wage is a function of both the outside and inside options when it is known that either the union or the firm will have to terminate the conflict when its strike funds are exhausted.

more centralized the negotiations over the base agreement, the more details must be left to be settled in subsequent local bargaining. At the local level, however, talks over implementation of the central agreement easily blend together with bargaining over additional wage increases. Local unions have bargaining power and it is unreasonable to expect such power to remain unused. In fact, wage increases above the central agreement, or wage drift, have comprised from one- to two-thirds of total wage growth in the Nordic countries since 1970 (Flanagan 1990). In Norway, wage drift has been as high as 80 per cent of aggregate wage growth in recent years (Rødseth & Holden 1990). Since local bargainers get the last word in the sense that local negotiations occur after the central agreement is settled, the actual degree of centralization attained in the Nordic countries, or anywhere else, is unclear. Does a central agreement imply central control over the total wage increase? The answer depends on whether or not industrial action is restricted at the local level.

To demonstrate this, consider local bargaining in a two-tiered system. The centrally negotiated wage, denoted q, is settled first and taken as given in local negotiations. Let d be wage drift, or wage increases obtained in local bargaining. The final wage is then $w = q + d$.

We examine first the case where strikes are allowed in local bargaining. Striking workers are not paid by the firm, of course. Although striking workers generally receive strike benefits, the benefits come from their own funds unless the strike is subsidized from outside. Central confederations that provide strike support do not allow locals to draw upon the funds whenever they wish. Thus union locals that are free to strike must supply their own funds. Here we assume that neither side receives outside support during a strike. Thus, the pay-offs when strikes are permitted are

$$\pi = \begin{cases} R - (q + d)L & \text{if there is an agreement} \\ 0 & \text{if there is a strike} \end{cases} \tag{7}$$

for the firm, and

$$u = \begin{cases} q + d & \text{if there is an agreement} \\ 0 & \text{if there is a strike} \end{cases} \tag{8}$$

for the union. Applying the bargaining solution (2), we have

$$d = \alpha(R/L) - q \tag{9}$$

as the expression for drift. The final wage is independent of the centrally negotiated wage (as long as all agreements including $d<0$ are possible). Smaller increases at the central level are offset exactly by larger increases at the local level. Central negotiations, in such a system, are a ritual without real impact on the economy.

However, this is not an accurate description of centralized bargaining in either Sweden or Norway. In both countries, the main agreement between the unions and the employers' association contains an industrial peace clause that forbids unions from calling strikes or go-slow actions (and forbids employers from calling lock-outs) as long as the central agreement is in force. This does not mean that locals have no credible threats in local bargaining. Workers may engage in work-to-rule actions where they follow work instructions in a pedantic way, decline to work overtime, and generally refuse to co-operate with the firm. The work environment legislation of the 1970s gave local unions new powers to disrupt production by refusing to overlook minor infractions of the law. In addition, the increased use of autonomous work groups gives workers greater control over production and hence more ways of reducing productivity without ceasing to work, at least nominally. Employers are generally unable to take locals to court for breaking the peace clause, or to reduce workers' pay during such actions, because it is difficult to prove that the peace clause has been broken.[5]

During a work-to-rule action, workers receive the centrally negotiated wage q while firms suffer a loss of output. We will assume that work-to-rule actions reduce output by the proportion θ, where $0<\theta<1$ (Moene 1988). Then the pay-offs in local bargaining under an industrial peace constraint are

$$\pi = \begin{cases} R-(q+d)L & \text{if there is an agreement} \\ \theta R - qL & \text{if there is a conflict} \end{cases} \tag{10}$$

for the firm, and

$$u = \begin{cases} q+d & \text{if there is an agreement} \\ q & \text{if there is a conflict} \end{cases} \tag{11}$$

[5] See Moene 1988 for an investigation of other forms of industrial action at the local level. For other models of wage drift, see Holmlund 1986; Holden 1988; 1989; 1990; Holmlund & Skedinger 1990; and Andersen & Risager 1990.

for the union. Substituting these pay-offs in equation (2), we get

$$d = \alpha(1 - \theta)(R/L) \tag{12}$$

as the outcome of local bargaining with a work-to-rule threat.

According to equation (12), drift is independent of the centrally set wage. Every increase or decrease of the centrally negotiated wage is passed on to the final wage (Holden 1989). When strikes are forbidden at the local level, wage drift adds a constant sum to whatever is obtained in central negotiations. The main impact of local bargaining in this case is to set a floor on wage growth to the extent that the centrally negotiated wage growth cannot be negative.[6] In two-tiered bargaining systems, the degree of centralization depends on the extent to which industrial action at the local level is restricted once the central agreement is signed. Our central result here is that if central negotiators prefer a lower wage than would result from local bargaining for any of the reasons elaborated in Chapter 11, central bargainers can restrain overall wage growth by negotiating a smaller increase at the central level provided the central bargain is backed up by an industrial peace clause.

(c) The Frequency of Industrial Conflict

One of the striking conclusions of Rubinstein's bargaining model is that the equilibrium is efficient in the sense that nothing is lost through conflict. Although the division of the pie is determined by the relative costs of delay, the equilibrium strategies entail an acceptance of the first offer that is made. This seems to leave the occurrence of strikes or lock-outs to random mistakes or

[6] The implication of (12) that drift is independent of the centrally negotiated wage increase is supported by econometric studies of Norwegian data by Steinar Holden (1989; 1990) and Asbjørn Rødseth and Holden (1990). The conclusions of studies of drift in other Nordic countries are more mixed. Robert Flanagan (1990) finds that drift is independent of the centrally negotiated wage in Norway, Sweden, and Denmark but not in Finland. Tyrväinen (1989) also concludes that drift in Finland partially offsets central wage increases. Bertil Holmlund and Per Skedinger (1990) and Douglas Hibbs and Håkan Locking (1991) find that drift partially offsets centrally negotiated wage increases in Sweden as well, but their study is confined to a single industry. All studies agree that the central agreement does influence the final wage in that drift does not fully offset centrally negotiated increases.

deviations from purely rational behaviour. Indeed, John Hicks (1963) argued that no theory of bargaining founded on rational behaviour with a unique solution could ever explain strikes, since both sides could then predict the outcome and agree to it without a costly conflict.

Yet the conclusion that industrial conflict is essentially random is belied by the fact that the frequency of strikes appears to follow predictable patterns (Kennan 1986). One of the most striking empirical regularities is the strong negative correlation between industrial conflict and the centralization of bargaining, as the following quote from Hibbs (1976: 1049) indicates:

> Simple calculation of strike volume for each type of bargaining system leaves no doubt that during the postwar period the average level of strike activity covaried with the degree of centralization: mean man-days lost per 1000 wage and salary workers are 425, 172 and 67 for decentralized, centralized and highly centralized systems, respectively.

Moreover, the effect of centralization on strike frequency can be observed over time within single countries as well as cross-nationally. Norway and Sweden were among the world's most strike and lock-out-prone countries during the inter-war years before collective bargaining was centralized. In the post-war period of centralized bargaining, in contrast, the frequency of industrial conflict in Norway and Sweden was among the lowest observed anywhere (Ingham 1974). With the recent decentralization of bargaining in Sweden, the frequency of strikes has risen again.

The usual way out of the Hicks paradox is to expand the bargaining model to include private information held by one side or both.[7] The most plausible candidate is the information that each firm gathers about the demand for its output. The difficulty that private information creates is easy to understand. Suppose the firm is hit by a sudden decline in demand. If the decline in demand was common knowledge, the union would adjust its expectations accordingly and contract negotiations would be no harder than usual. But if the firm notifies the union that conditions have worsened, will the firm be believed? After all, the union knows that it is in the firm's interest to say that con-

[7] The theoretical and empirical literature on strikes is reviewed in Kennan 1986.

ditions have worsened, even if they haven't. Whether demand is falling or rising, the firm always has an incentive to be pessimistic in its message to the union. Knowing this, the union discounts any message from the firm that is not costly for the firm to transmit. One mechanism whereby firms might credibly communicate a worsening of conditions is to lay workers off. Another way is to endure a strike. In fact, the empirical evidence indicates that lay-offs and strikes are substitutes at the firm level in the sense that strikes (in the USA) are procyclical (Kennan 1986). One can speculate that lay-offs are generally used to communicate during downturns in demand. Strikes are more likely to occur during expansions when the union suspects that conditions are better than the firm says they are.

This leads to the following simple explanation of the relationship between centralization and industrial conflict. There is a clear asymmetry in the information available to a firm, and the information held by the union. The existence of an asymmetry in the information held by an association of employers at the industry level and an industrial union is less obvious. The union can do its own studies of the demand for an industry's output. At the national level, the existence of any asymmetry of information is even less likely. The national confederation of trade unions has access to the same information about the state of the aggregate economy as the national confederation of employers. In Norway, for instance, both sides receive the same government reports prepared by the Bureau of Statistics. As a consequence, centralized bargaining rarely fails to come to agreement without conflict.

(d) Local Bargaining as Revenue-Sharing: The Choice of Effort

All wage bargaining entails a sort of profit-sharing. The higher the firm's profits, the more the union is able to take out in wages. When profits are low, unions must settle for lower wage growth or lower employment (or a combination of the two). At the local level, wage bargaining is a form of profit-sharing between a firm and its work-force. At higher levels, the profits that are shared are aggregated over an industry or an entire economy. Unless a firm is large relative to the bargaining unit, the wage contract will not be sensitive to *its* profits. Only at the local level, therefore, will the implicit profit-sharing affect the firm's and unions' decisions regarding variables outside the wage agreement. Three variables seem particularly important: workers' effort on the job, employment, and investment.

We start with workers' effort. It is often argued that profit-sharing has a negligible impact on individual incentives to work harder in all but the smallest plants.[8] Some aspects of work effort, however, are decided collectively. This is particularly true of the adoption of new techniques that increase productivity but demand greater effort on the part of the work-force. It is the part of effort that is collectively determined that we are interested in here.

Following the efficiency wage framework, we assume that labour input can be written as eL, where e is the efficiency of labour: $R = R(eL)$ with $R'(eL) > 0$ and $R''(eL) < 0$. For simplicity, we assume that employment is fixed with $L = 1$. Rather than assume e is determined by relative wages adjusted for employment, as we did before, here we consider e to be a matter of choice. Workers, we assume, care about both wages and their effort. Beyond some level of effort, work is unpleasant. Let effort be measured such that $e = 1$ is the level of effort that workers will expend without requiring compensation. Without loss of generality, then, we can limit our attention to $e \geqslant 1$.

For convenience we adopt the particularly simple specification of workers' pay-offs of

$$u(w, e) = \begin{cases} w - v(e) & \text{if there is an agreement} \\ 0 & \text{if there is a strike} \end{cases} \tag{13}$$

where $v(1) = 0$, $v'(e) > 0$ and $v''(e) \geqslant 0$ for $e > 1$. Again we assume there is no outside strike support. Note that striking workers lose w but save the disutility of effort $v(e)$. Profits are given by $\pi = R(e) - w$ when an agreement is reached and zero in the case of conflict. To ensure an internal solution, we assume that $R'(1) > v'(1)$. Applying the bargaining solution of equation (2), we have

[8] This is frequently referred to as the $1/n$ problem. See the essays collected in Blinder 1990 for discussions and empirical tests of the relationship between profit-sharing and productivity.

$$w = \alpha R(e) + (1 - \alpha)v(e) \tag{14}$$

as the expression for the wage in the case of decentralized bargaining. Incorporating (14), the pay-offs for the union and the firm upon signing the labour contract with local bargaining can be written as

$$u = \alpha[R(e) - v(e)] \tag{15}$$

$$\pi = (1 - \alpha)[R(e) - v(e)]. \tag{16}$$

These last two equations display the similarity of local wage-bargaining and profit-sharing.

There are three plausible assumptions that can be made about the choice of effort. The first is that effort, in the sense of new techniques or the reorganization of work, is bargained over at the local level. The other two alternatives are that effort is set unilaterally by either the local union or the firm. As can be easily seen from (15) and (16), all three assumptions result in the same first-order condition for effort:

$$R'(e) = v'(e). \tag{17}$$

Effort is set at the collectively optimal level where the marginal increase in revenue equals the cost whether effort is set by the union, the firm, or through bargaining. With local bargaining, the union internalizes the full costs and benefits of effort, and so does the firm.[9]

In the case of centralized bargaining, we assume that the employers' association seeks to maximize aggregate profits while union pay-offs are unchanged. Assuming there are n firms, the same wage equation (2) specifies a tariff wage of

$$q = \alpha\frac{1}{n}\sum R(e) + (1 - \alpha)v(e). \tag{18}$$

The tariff wage reflects the average productivity of the entire sector (or economy). If the firm is small relative to the bargaining unit, the tariff wage is exogenous from the point of view of local bargainers.

[9] This strong result depends critically on the strong assump workers' preferences. It does not hold for more general preferences over wages and effort.

In this case there is maximal disagreement over effort between employers and workers. If the local union controls effort, effort would set close to the minimal level $e = 1$. The benefits of greater effort are shared by all through centralized bargaining while the costs are borne by local workers alone. The result is suboptimal effort and reduced welfare for all. In contrast, employers, if they could, would increase effort as much as possible subject to the constraint that the firm be able to attract sufficient labour. Only if the level of effort is specified in the labour contract can the optimal effort be obtained with centralized bargaining. If bargaining is centralized at the industry level, effort may be bargained over to some extent. But centralization at the national level makes bargaining over effort impossible beyond the setting of minimal standards that apply in all industries.

Yet centralized bargaining on the national level is typically accompanied by supplementary bargaining on the local level. The model of two-tiered bargaining is the relevant model for unions that receive drift in addition to the central wage. We consider the case of two-tiered bargaining when strikes at the local level are prohibited. In the present context, it is natural to model the work-to-rule action as equivalent to working with minimal effort, or $e = 1$. Thus, for the firm, we have the pay-offs

$$\pi = \begin{cases} R(e) - (q + d) & \text{if there is an agreement} \\ R(1) - q & \text{if there is a conflict.} \end{cases} \tag{19}$$

For the union we have

$$u = \begin{cases} q + d - v(e) & \text{if there is an agreement} \\ q & \text{if there is a conflict.} \end{cases} \tag{20}$$

In this case, local bargaining results in wage drift of

$$d = \alpha[R(e) - R(1)] + (1 - \alpha)v(e). \tag{21}$$

Given a tariff wage of q and drift given by (21), profits are equal to

$$\pi = (1 - \alpha)[R(e) - v(e)] + \alpha R(1) - q \tag{22}$$

while the union receives

$$u = \alpha[R(e) - v(e)] - \alpha R(1) + q. \tag{23}$$

As can be seen immediately, effort will be set at the collectively optimal level regardless of who chooses, exactly as in the case with firm-level bargaining.

Thus, mixed bargaining systems, unlike purely centralized bargaining systems, do not distort the decision over the level of effort. Local bargaining, even if conducted under a peace clause, gives workers a stake in the performance of their firm and thus increases workers' willingness to accept higher effort. At the same time, the more effort workers exert, the greater the threat of withdrawing co-operation during a work-to-rule action. In this way, the employer shares the cost of effort, as well as the benefit. The importance of local bargaining in providing a reward for greater effort on the job is a strong argument against proposals to cap or eliminate drift in centralized bargaining systems.

(e) Local Bargaining as Revenue-Sharing: Employment and Investment

Local bargaining as a form of profit-sharing also affects decisions regarding employment and investment. To examine bargaining structure and employment, we consider a model with a fixed number of identical firms in which the capital stock and work effort are given. Revenue depends on employment: $R = R(L)$ with $R(L) \geqslant 0$, $R'(L) > 0$ and $R''(L) < 0$. Unions continue to maximize their wage minus the disutility of work $u = w - v$ where v is now a positive constant. Firms, as always, maximize profits.

The traditional right-to-manage model says that firms choose employment along the demand-for-labour curve where profits are maximized for a given wage, or:

$$R'(L) = w. \tag{24}$$

Equation (24) is appropriate when the firm is small compared to the bargaining unit. In that case, each firm considers the wage to be exogenous and optimally adjusts employment. But if the firm is large in relation to the bargaining unit, as is the case with decentralized bargaining, then firms might not ignore the way that current employment influences future wage bargains.

According to our standard equation for local negotiations, the wage is given by

$$w = \alpha \frac{R(L)}{L} + (1-\alpha)v \tag{25}$$

as long as the lower bound is not binding. With wages set according to (25), profits are

$$\pi = (1-\alpha)[R(L) - vL]. \tag{26}$$

If employers choose L to maximize (26), they would set employment according to the condition

$$R'(L) = v < w. \tag{27}$$

With local bargaining, employers can lower the wage by raising employment and thereby lowering output per worker. Since v may well be less than the competitive wage, local bargaining can lead to a full employment, suction equilibrium where employers' desire to expand is constrained by the supply of labour similar to the equilibrium of Weitzman's (1983; 1984) share economy.[10] Local unions may have sufficient power to block expansions of employment that reduce their wages, but at least employers would desire to hire more workers with local wage-bargaining than with centralized bargaining.

Equation (27) implies, however, that firms are employing more workers than they would like at their current wage. It is often argued that this is not an equilibrium in that firms could increase profits by laying off workers and returning to their demand-for-labour curve as soon as the wage contract is signed and wages are fixed. What this argument ignores is that there will be new negotiations in one or two years. If the firm cannot suddenly expand its work-force just before the next round of bargaining begins, the wage in the future will be influenced by the level of employment chosen in the present.

This can be presented with a simple model by writing the intertemporal problem facing the firm as

[10] The equivalence of local wage-bargaining and profit-sharing is investigated in greater detail in Moene 1986. See Raaum 1990 for a similar conclusion regarding local bargaining and employment in a model where wages are set unilaterally by the union.

$$\max_{L_t} V = \sum_{t=0}^{\infty} \delta^t [R(L_t) - w_t L_t i] \quad \text{with} \quad \delta \in [0,1]. \tag{28}$$

We assume that the wage is set in a collective agreement negotiated at the beginning of every period. We assume, in addition, that the level of employment can only be altered once each period, immediately after the wage has been set. Thus w_t is fixed when L_t is chosen, but the choice of L_t affects w_{t+1}:

$$w_{t+1} = \alpha \frac{R(L_t)}{L_t} + (1-\alpha)v. \tag{29}$$

With w_t determined by (29), the first order condition for L_t is

$$R'(L_t) - w_t - \alpha\delta \frac{L_{t+1}}{L_t}\left[R'(L_t) - \frac{R(L_t)}{L_t}\right] = 0 \tag{30}$$

If we assume a steady state with $L_t = L_{t+1}$ for all t, equations (29) and (30) together imply

$$R'(L) = \left[\frac{\alpha(1-\delta)}{1-\alpha\delta}\right]\frac{R(L)}{L} + \left[\frac{1-\alpha}{1-\alpha\delta}\right]v. \tag{31}$$

If $\delta = 0$, then equation (31) reduces to the (24) where the firm chooses a point on the demand-for-labour curve. The more firms care about the future, i.e. the higher is δ, the higher the firms' preferred level of employment. If $\delta = 1$, equation (31) reduces to (27).

According to this model, decentralized wage-setting is equivalent to centralized wage-setting only when firms have an extremely short time horizon. Otherwise, local bargaining increases employment.[11] In this way, the debate over whether or not employment is covered in the labour agreement or set by the firm that has occupied so much of the literature has been misguided. When bargaining is centralized at the national level, agreements covering employment are infeasible. Even at the industry level, agreements over manning rules and the like are difficult if work practices differ among plants. Thus with national wage contracts

[11] Hoel 1989*b* examines this question in greater detail in a model where the firm can change employment at any time, but it must pay hiring and firing costs. In his model, local bargaining induces firms to hire more workers than centralized bargaining when marginal hiring costs go up sharply as the number of workers who are hired increases.

and, we suspect, with most industry-level contracts, firms set employment taking the union wage as given. When bargaining is decentralized, in contrast, employment may be set off by the demand-for-labour curve, whether or not employment is set by the firm or covered indirectly by negotiations over work rules and the like. What matters fundamentally is the level of bargaining, not the coverage of the labour agreement.

We still have to analyse the impact of two-tiered bargaining on employment. From equation (12), we have $d = \alpha(1-\theta)(R/L)$ as the expression for drift. Profits, then, are equal to

$$\pi = R(L) - (q+d)L = [1 - \alpha(1-\theta)]R(L) - qL \tag{32}$$

from which we get

$$R'(L) = \frac{q}{1-\alpha(1-\theta)} \tag{33}$$

as the first order condition for employment. From equation (33) it is not obvious how to compare the firms' demand for labour under two-tiered bargaining as opposed to either decentralized or purely centralized bargaining. In fact, the mixed bargaining case produces a demand for labour that is in between the two pure cases.

To demonstrate this, we need to show that the right-hand-side of (33) is greater than v but less than $w = q + v$. Under the assumptions of this subsection, the final wage is the same whatever the level of bargaining. Therefore, we can use equation (25) to write

$$q = w - d = \alpha\frac{R(L)}{L} + (1-\alpha)v - d = \alpha\theta\frac{R(L)}{L} + (1-\alpha)v. \tag{34}$$

From (34) it is straightforward to calculate that $q[1-\alpha(1-\theta)]^{-1} > v$. To see that $q[1-\alpha(1-\theta)]^{-1} < q + v$, note that $q + d < (R/L)$ or

$$q < [1-\alpha(1-\theta)]\frac{R(L)}{L} \tag{35}$$

or

$$\left[\frac{\alpha(1-\theta)}{1-\alpha(1-\theta)}\right]q = \left[\frac{1}{1-\alpha(1-\theta)} - 1\right]q < a(1-\theta)\frac{R(L)}{L} = d. \tag{36}$$

Thus, equations (34) and (36) imply

$$v < \frac{q}{1 - \alpha(1 - \theta)} < q + d = w. \tag{37}$$

For the determination of effort, two-tiered bargaining was equivalent to purely local bargaining. This is not the case for the demand-for-labour. Two-tiered bargaining results in a demand for labour that is less than the labour demand with purely local bargaining but more than the demand for labour with purely centralized bargaining.

A third important aspect of the performance of centralized versus decentralized bargaining is the impact of collective bargaining on investment. Here the standard results are exactly the opposite of what we found in the case of employment (Grout 1983; Hoel 1990; Moene 1990). Investment in fixed capital increases the cost to the firm of a work stoppage and therefore increases the union's bargaining power. Since, with local bargaining, firms know that greater fixed costs increase their vulnerability to the threat of a strike, firms invest less.

Let us add capital (K) to the model, with $R = R(K, L)$ and $C(K)$ as the cost of capital. All investment we assume is fixed in the sense that once the capital is installed, it has no other use. A strike or lock-out stops production, but it does not eliminate the cost of the capital equipment. Formally, with fixed investment we have

$$\pi = \begin{cases} R(K, L) - wL - C(K) & \text{if there is an agreement} \\ -C(K) & \text{if there is a strike.} \end{cases} \tag{38}$$

With decentralized bargaining, the wage is given by $w = \alpha(R/L) + (1 - \alpha)v$, as before, assuming the union is sufficiently powerful to raise the wage above the competitive level. Thus profits equal

$$\pi = (1 - \alpha)[R(L, K) - vL] - C(K) \tag{39}$$

upon conclusion of the wage agreement. It is apparent from (39) that local bargaining raises the implicit cost of capital by the multiple $(1 - \alpha)^{-1}$, holding employment constant. In contrast, centralized bargaining does not raise the implicit cost of capital to the firm, in so far as the wage agreement is independent of any one firm's investment decisions.

One cannot conclude that local bargaining will reduce investment, however, because local bargaining may increase employment which raises the productivity of capital. Whether local bargaining results in more or less investment than centralized bargaining depends on such aspects of the environment as the industry's demand curve and the supply constraints for capital and labour inputs. The most that can be said that is generally true is that the capital–labour ratio is lower with local bargaining than with centralized bargaining since decentralization lowers the implicit cost of labour and raises the implicit cost of capital. These issues are pursued further in the next section in a model where investment is studied in the form of the entrance of new firms or the building of new plants.

(f) Entry and Exit

Until now, we have assumed that the number of firms (or plants) was fixed and that all firms shared the same technology. Yet much of the dynamic of capitalist economies is due to the continual entrance of new firms and the failure of existing firms. Expansions are marked by the building of new plants; contractions by the closure of old ones. Entry and exit alter more than the quantity of labour and capital employed. New entrants often bring new techniques, while departing firms leave behind the most efficient. In this way, both entry and exit change the mix of firms in the industry and increase average productivity. In this section, we investigate the way in which the pace of both entry and exit is affected by the level of bargaining.

In order to capture the effect of entry and exit on productivity and average wages, we need a model with heterogeneous firms. The very simplest such model consists of just two types: high-productivity and low-productivity firms, denoted by subscripts H and L respectively. For simplicity, we assume that all changes in capital and labour employed are due to entry or exit. Employment per plant of either type is fixed at $L = 1$. Non-labour costs of production C_i, which may differ between types, are also fixed at the plant level.[12] Profits are given in each type of firm by

[12] Many of these assumptions can be relaxed without altering the results. Moene and Wallerstein (1992) consider a model with variable employment within each plant. The model examined in Moene and Wallerstein 1991*a* contains an infinite number of firm types, each corresponding to a point on a line segment.

$$\pi_i = p\beta_i - w_i - C_i, \quad \text{for} \quad i = H, L, \quad \text{with} \quad \beta_H > \beta_L. \tag{40}$$

The term β_i is the quantity produced by a plant of type i. Let the number of high-productive firms be n_H and the number of low-productive firms be n_L. Throughout we assume that the price is independent of the output of any individual firm but dependent on the aggregate output of the $n_H + n_L$ firms:

$$p = p(n_H\beta_H + n_L\beta_L) \quad \text{with} \quad p'(\cdot) < 0. \tag{41}$$

In comparing different levels of bargaining, it is useful to use a competitive labour market as a benchmark. In the competitive case, all employers pay the same wage

$$w_H = w_L = r \tag{42}$$

where r is the lowest wage that employers can pay and still attract sufficient labour. Since the disutility of effort plays no role in this section, we set $v = 0$ to simplify the notation. Assuming that firms must continue to pay the costs C_i during a labour conflict, local wage-bargaining produces a wage of

$$w_i^{LB} = \max(\alpha p\beta_i, r). \tag{43}$$

Here we include workers' outside option as a lower bound on possible wage settlements.[13] Note that with local bargaining, more productive firms pay higher wages, assuming that $w_H > r$.

We assume that wage-bargaining at the industry level sets a uniform wage for all firms. If negotiators for employers seek to maximize total profits in the industry, industry-level bargaining produces a wage of

$$w_H^{IB} = w_L^{IB} = w^{IB} = \max\left(\alpha p \frac{n_H\beta_H + n_L\beta_L}{n_H + n_L}, r\right). \tag{44}$$

We assume throughout that the union at the industry level is powerful enough to affect the wage, or $w^{IB} > r$.

Equations (40) to (44) can be used to represent several dif-

[13] There is also an upper bound for the wage given by the constraint that $\pi \geqslant 0$. Implicit in (43) is the assumption that if one starts with a sufficiently high price such that $w_i = \alpha p\beta_i$, and then lets the price fall, the lower bound binds before the upper bound. This is equivalent to assuming that $r > (\alpha/(1-\alpha))C_i$. If $r < (\alpha/(1-\alpha))C_i$, the upper bound binds first and workers' wage would be given by $w_i = \min(\alpha p\beta_i,\ p\beta_i - C_i)$. The choice is arbitrary and inconsequential for the results.

ferent kinds of industrial structure. In the first kind we consider, the supply of high productivity firms is limited. The size of the industry is determined by entry and exit of less productive firms. Thus, n_H is fixed and n_L is endogenous. We assume that the more-productive firms are more profitable than the less-productive firms, or $p\beta_H - C_H > p\beta_L - C_L$. Since, as long as this condition is satisfied, nothing depends on $C_H \neq C_L$, we might as well let $C_H = C_L = C$. Second, we assume that demand is high enough relative to the supply of more-efficient firms that some less-efficient firms can profitably enter in a competitive labour market, or $p(n_H\beta_H)\beta_L - r - C > 0$. Third, we assume that the potential supply of less-productive firms is greater than demand, so that free entry implies that the profits of the less-efficient firms are driven to zero, or

$$p(n_H\beta_H + n_L\beta_L)\beta_L - w_L - C = 0. \tag{45}$$

The more-productive firms, within this industry structure, receive Ricardian rents.

This case could be interpreted as an industry with a mature technology and some industry-specific factor of production that cannot be expanded without a reduction in quality. Such a factor of production could be a rich vein of ore, an advantageous location, superior skills acquired through 'learning by doing,' managerial expertise, or simply a successful company culture. Alternatively, this model can be viewed as a model of a declining industry in which capacity exceeds demand. As demand shrinks, the least-productive firms are the first to close.

The comparison of a competitive labour market, local bargaining, and industry-wide wage-bargaining can be summarized as follows:

$$w_H^{LB} \gtrless w^{IB}, \tag{46}$$

$$w^{IB} > w_L^{LB} = r, \tag{47}$$

$$p^{IB} > p^{LB} = p^C, \tag{48}$$

$$n_L^{IB} < n_L^{LB} = n_L^C, \tag{49}$$

where the superscripts IB, LB, and C represent industry bargaining, local bargaining, and the competitive case respectively. That $w^{IB} > r$ is an assumption. That $w_L^{LB} = r$ follows from free entry of less-productive firms. If less-productive firms could pay wages above r, they would be earning positive profits.[14] Entry of additional firms would then drive the price down until both the constraint $w_L^{LB} \geqslant r$ and the constraint $\pi_L \geqslant 0$ are binding. Equations (48) and (49) follow from the zero-profit condition for less-productive firms. From equation (45) we know that $(dp/dw_L) = (1/\beta_L) > 0$. Thus, $p^{LB} = p^C$ since $w_L^{LB} = r$ and $p^{IB} > p^{LB}$ since $w^{IB} > w_L^{LB}$. The negative relationship between price and total output implies that the number of less-efficient firms must decline as the price increases.

In general we cannot tell whether w^{IB} is higher or lower than w_H^{LB}. Both cases may apply depending on the bargaining power of the union. If the union is weak (if α is low enough) industry-wide bargaining leads to a wage close to r and a price close to p^C. In this case (43) and (44) imply that $w_H^{LB} > w^{IB}$ since the average labour productivity is higher in the high productivity firms than in the industry as a whole. If the union is strong (if α is high enough), industry-wide bargaining produces a wage sufficiently high to keep all less-productive firms out of the market. In this case (43) and (44) imply that $w_H^{LB} < w^{IB}$ since the average productivity is the same in the two cases while the price p is higher with industry bargaining.

With a fixed supply of more-efficient firms, the price is determined by the zero-profit condition for the less-efficient firms. The more-efficient firms receive Ricardian rents. In this environment, industry-level bargaining has the textbook effects of raising the price, reducing employment, and increasing average productivity by driving some of the less-productive firms out of the market. The union pushes the industry up its demand curve and captures some of the monopoly rents. Note that the more-efficient firms also obtain a share of the monopoly rents as $d\pi_H/dw = (\beta_H/\beta_L) - 1 > 0$. With both employers and employees in the less efficient firms receiving their outside option, the gains from industry-level bargaining are paid for by consumers.

In contrast, local bargaining is very similar in its effects to a competitive labour market. Free entry drives the union wage

[14] This statement follows from the assumption that the constraint $w_L \geqslant r$ binds before the constraint $\pi_L \geqslant 0$ binds. If the zero profit constraint binds first, then the results are the same provided one assumes that less-efficient firms continue to enter as long as $\pi_L \geqslant 0$.

down to the competitive wage in the less-efficient firms. Since the less-efficient producers determine the price, no monopoly rents result from wage bargaining. The only difference unionization makes is that workers in more-efficient firms are able to obtain a share of the Ricardian rents. The only losers from local bargaining are the owners of the more-efficient firms who are forced to share the rents with their workers.

An alternative, equally interesting industry structure can be represented by the opposite assumption that the number of less efficient firms n_L is fixed while the number of more efficient firms n_H is endogenous. The interpretation of this case is that new state-of-the-art plants embodying the latest technological advances are more efficient than plants built in the past. The assumption that the entrants would be the most productive seems appropriate for a growing industry with a developing technology.

To obtain an equilibrium with both types of firms in the market, we must have $C_H > C_L$. This last condition can be justified by the nature of investment in new plants. Let $C_L = C$ and $C_H = C + I$ where I is the cost of new investment. Once a plant is built and equipped, the cost I is sunk. Thus firms will continue to operate existing plants as long as revenues cover the labour and non-labour operating costs $w + C$. Before building, however, firms will not invest unless revenues will cover all costs $w + C + I$.[15] Firms that enter earn quasi-rents on their sunk costs. Free entry implies, however, that the more-productive firms earn zero profits *ex ante*:

$$p(n_H\beta_H + n_L\beta_L)\beta_H - w_H - (C + I) = 0. \tag{50}$$

In the case of an expanding industry with embodied technical change, the comparison of a competitive labour market, firm-level bargaining and industry-level bargaining can be summarized by the equations:

$$w_H^{LB} > w^{IB} > r, \tag{51}$$

[15] This is a simplified static representation of a necessarily dynamic story. See Moene and Wallerstein 1991*b* for an explicitly dynamic version in which both the number of less-efficient types and the number of more-efficient types are endogenous.

(a) Conflict over Local Bargaining Rights

One of the perennial conflicts in centralized bargaining systems concerns local bargaining rights. Workers may care about other workers' drift because they care about other workers' final wage for all the reasons discussed in Chapter 11. In addition, workers care about others' drift because other workers' drift affects the central agreement and their own final wage. To show this, we continue with the last model of an industry with two types of firms: high-productivity firms with output of β_H and low-productivity firms that produce β_L. Workers' final wage is equal to the centrally negotiated wage q, which is the same for all, and drift d_i, which varies with each firm's productivity. We assume that the negotiators for the employers seek to maximize total industry profits, while union negotiators at the central level seek to maximize the average wage. Formally, we can characterize industry-level bargaining by the pay-offs

$$\pi = \begin{cases} \Sigma n_i[p\beta_i - (q + d_i) - C_i] & \text{if there is an agreement,} \\ -\Sigma n_i C_i & \text{if there is a strike} \end{cases} \tag{1}$$

for employers and

$$u = \begin{cases} q + (\Sigma n_i d_i / \Sigma n_i) & \text{if there is an agreement,} \\ 0 & \text{if there is a strike} \end{cases} \tag{2}$$

for the union. Equations (1) and (2) imply that central negotiators anticipate the outcome of subsequent bargaining at the local level when setting the base wage.

The outcome of central bargaining, according to our standard formula, is the tariff wage

$$q = \alpha p\left(\frac{n_H\beta_H + n_L\beta_L}{n_H + n_L}\right) - \left(\frac{n_H d_H + n_L d_L}{n_H + n_L}\right). \tag{3}$$

According to (3), the tariff wage consists of the share α of the industry's average revenues minus the average anticipated drift. Increases in drift reduce the tariff wage. If we consider the final wage of workers in the sector j, where j can be either H or L and $k \neq j$, we have

$$w_j = \alpha p\left(\frac{\Sigma n_i\beta_i}{\Sigma n_i}\right) + \left(\frac{n_k}{\Sigma n_i}\right)(d_j - d_k). \tag{4}$$

The wage for workers in each sector is an increasing function of their own drift and a decreasing function of the drift received by workers in the other sector, holding n_H and n_L constant. Moreover, if we consider matters from the point of view of an individual local union, this effect is amplified. Each group of workers receives the full benefit of an increase in their drift, while the cost in terms of a lower tariff wage is borne by all. From the point of view of an individual local, increasing its wage drift as much as possible is a dominant strategy. If output in local conflicts can be reduced from β_i to $\theta\beta_i$, we have as before that $d_i = (1-\theta)\alpha p\beta_i$. If θ was chosen by each local independently, all locals would choose $\theta = 0$ and bargaining would be fully decentralized.

There are numerous externalities, many of which are discussed in Chapter 11, that can be invoked to argue that unions are caught in an n-person prisoners' dilemma where the stable decentralized solution is Pareto inferior. Holden and Raaum (1989) characterize centralization as a Pareto optimal equilibrium of an iterated n-person prisoners' dilemma supported by trigger strategies. Yet a simpler solution seems more realistic. The unions can sign a legally binding collective agreement. The feasibility of making binding commitments does not seem to be an important issue.

Centralized bargaining, therefore, entails a collective choice of $\theta > 0$ for all locals enforced by labour courts. But a common constraint on all only eliminates the conflict over local bargaining rights when firms are equally productive. If firms differ in terms of productivity, conflicts of interest within the union over local bargaining rights are still present. Let there be a restriction on legal industrial action at the local level for all, formalized by the assumption that output in local conflicts can only be reduced by a common value of θ such that $d_i = (1-\theta)\alpha p\beta_i$. Inserting this in the expression for the total wage $q + d_j = w_j$, $j = H, L$, given by (4) we obtain

$$w_j = \frac{\alpha p}{\Sigma n_i}[\Sigma n_i\beta_i + n_k(1-\theta)(\beta_j - \beta_k)] \tag{5}$$

assuming that $w_j \geqslant r$. Holding the price of output constant, workers in the high-productivity sector benefit from a relaxation of the constraints on local bargaining rights or a lower θ since

$\beta_H - \beta_L > 0$. For workers in the low-productivity sector, a lower θ leads to a lower final wage. Workers in low-productivity firms do best when drift is disallowed altogether or $\theta = 1$.

This conflict between workers in high-productivity and low-productivity firms may be attenuated in industry bargaining by the effect of θ on the price of output. If the domestic industry faces a downward sloping demand curve, then the union can increase the price of output and create monopoly rents by lowering θ in the case of an expanding industry or raising θ in the case of a declining industry. The higher price of output, a benefit for all workers who remain in the industry, may or may not outweigh the distributional effect on those potentially harmed by higher or lower drift. The lower the elasticity of demand, the sharper the conflict over local bargaining rights. Conflict over local bargaining rights is even sharper in bargaining systems that are centralized at the national level. If national level bargainers determine an optimal final wage, w^*, in line with the models of Chapter 11, then the central agreement will set the tariff wage to be the target wage minus average drift, or $q = w^* - (\Sigma n_i d_i / \Sigma n_i)$. When bargaining is centralized at the national level, conflict over local bargaining rights is a zero-sum game among workers.

(b) The Instability of Centralized Bargaining

The conflict over constraints on drift is only one of many conflicts that may exist among workers in centralized bargaining systems. There is conflict over the wage differential received by workers with high levels of education, or workers who work under harsh conditions, or workers who work in the private sector. With centralized bargaining, there are at least two bargains that must be struck. The explicit bargaining is between unions and employers. However, there is another bargain that must be concluded among the unions over the distribution of allowable wage increases. One might add a third bargain among employers. Unlike the bilateral bargaining between the unions and the employers' confederation, the bargaining within the unions (and among employers) is multilateral. The possibility of forming various coalitions creates instabilities that are not present in the bilateral case.

Let the outcome of bargaining among unions be characterized by a standard characteristic function with transferable utility, where $v(S)$ represents the total that could be obtained by the coalition S and $v(N)$ represents the total pay-off obtainable by the grand coalition of all unions. We assume that centralized bargaining is efficient in the strong sense that the characteristic function is strictly superadditive:

$$v(S_1) + v(S_2) < v(S_1 \cup S_2) \tag{6}$$

for all non-intersecting S_1 and S_2. Let the pay-offs to each union be denoted by x_i. The core is defined to be the set of pay-offs $x_1, x_2, \ldots, x_n$ such that

$$\sum_{i \in N} x_i = v(N), \quad \text{and} \tag{7}$$

$$\sum_{i \in S} x_i \geqslant v(S) \quad \text{for all } S \subset N. \tag{8}$$

Centralized bargaining can be efficient in the strong sense of equation (6), yet be unstable in the sense that no allocation satisfying equations (7) and (8) exists.

Suppose the core does exist, a yet stronger assumption. Then the wage bill specified in the central agreement can be allocated among unions in such a way that no subset of unions could do better by bargaining separately. Leif Johansen (1982) argued, however, that the core requires excessively acquiescive behaviour on the part of actors to be a realistic solution concept in many circumstances. An allocation of $v(N)$ is in the core as long as no subset of actors can do better by withdrawing. More typically, Johansen argued, actors demand what they could obtain outside the grand coalition plus a share of the surplus they create by joining.

Thus what we will call the Johansen core is defined to be an allocation of pay-offs to individuals $x_1, x_2, \ldots, x_n$ that satisfies

$$\sum_{i \in N} x_i = v(N), \quad \text{and} \tag{9}$$

$$\sum_{i \in S} x_i \geqslant v(S) + \lambda_S[v(N) - v(S) - v(N \backslash S)] \quad \text{for all } S \subset N. \tag{10}$$

According to (10), each group S demands what it could get outside the grand coalition, $v(S)$, plus the fraction λ_S of the surplus it brings to the coalition by joining, $[v(N) - v(S) - v(N \backslash S)]$. Note that the ordinary core is defined by equations (9) and (10) with $\lambda_S = 0$. It is clear that as the aggressiveness of the actors increases, that is as λ_S rises, the Johansen core may be reduced. Indeed, the Johansen core cannot possibly exist unless $\lambda_S + \lambda_{N \backslash S} \leqslant 1$. Otherwise the demands for shares of the surplus are incompatible.

How should the weights λ_S be determined? One natural way is to think of the λ_Ss as being the product of bargaining between the coalition S and its complement $(N \backslash S)$ over sharing the surplus. That is, coalition S threatens to leave the bargaining table and bargains with the remaining players over how to share the surplus should it remain. In this case, we have $\lambda_S + \lambda_{N \backslash S} = 1$.

Under these conditions, the Johansen core is almost always empty. Equations (9) and (10) imply

$$\sum_{i \in N \backslash S} x_i \leqslant v(N \backslash S) + (1 - \lambda_S)[v(N) - v(S) - v(N \backslash S)]. \quad (11)$$

But if $(1 - \lambda_S) = \lambda_{N \backslash S}$, then equations (10) and (11) imply that the inequality signs must be replaced by strict equality:

$$\sum_{i \in S} x_i = v(S) + \lambda_S[v(N) - v(S) - v(N \backslash S)] \quad \text{for all } S \subset N. \quad (12)$$

Equations (9) and (12) constitute a set of $2^n - 1$ equations to determine n variables $x_1, x_2, \ldots, x_n$. A solution will not exist except in very special circumstances. In general, it is impossible to obtain a centralized agreement that is 'renegotiation proof'. All possible allocations leave some group worse off than they could be if they withdrew from the grand coalition and bargained over the terms of rejoining.

Thus, even if all unions (or all employers) could obtain a higher level of welfare with centralized bargaining than they could by bargaining separately, it may still be impossible to distribute the gains from centralized bargaining in a way that maintains co-operation. Of course, the lack of a solution reflects the weakness of the theory of n-person bargaining rather than the impossibility of wage-setting at the national level.[1] National-level wage-bargaining did exist in Sweden for four decades, and it exists still in Norway and Finland, albeit in attenuated forms. If we have devoted much more space to the economic consequences of centralized bargaining than to the political conflicts over the level of bargaining, it is because the politics of bargaining are so poorly understood.

[1] There are many solution concepts in co-operative game theory that could be applied, but all suffer from the defect of assuming that the solution is Pareto optimal. Thus they are of little help in studying when Pareto optimal solutions are obtained and when they are not.

14
Conclusion

The literature on collective bargaining consists of a multiplicity of models, each with a different focus and, seemingly, a different conclusion. Robust conclusions that are not contradicted by some other plausible specification of the problem are difficult to find. In this respect, the literature on collective bargaining is similar to other topics in the field of industrial organization. This is discouraging, but it is better to recognize the diversity of results than to make claims as if economic theory had a clean, simple implication regarding the costs and benefits of different bargaining structures.

It is also disconcerting, after covering such a variety of topics, to list what has been left out. One of the most important is the effect of bargaining level on wage dispersion. The reduction of wage differentials is among the most visible effects of centralized bargaining in the Nordic countries (but not in Austria).[1] The economic effect of an egalitarian wage structure is highly controversial. Claims that larger wage differentials are needed to provide adequate individual incentives must be balanced against contrary claims that narrow wage differentials within the firm promote co-operation among workers and higher productivity.[2]

Also missing is a discussion of the level of bargaining and inflation.[3] In this essay we have followed the theoretical literature, though not the empirical literature, and focused exclusively on real models in which monetary policy has no role to play. This leads to another omission, namely the interaction of union wage-

[1] See Freeman 1988; Rowthorn 1989*a*; 1989*b*; and Kalleberg and Colbjørnsen 1990 for empirical studies of centralization and wage dispersion. Theoretical studies have been done by Byoung Heon Jun (1989); Wallerstein (1990); and Moene and Wallerstein (1991*c*).

[2] It should be noted that managers' preference for a less egalitarian wage structure is as self-serving as the preference of unions representing low-paid workers for a more egalitarian wage structure.

[3] For studies of the inflationary consequences of mixed bargaining systems, see Holden 1991*a*; 1991*b*; and Nymoen 1991.

setters and government policy-makers.[4] Implicit in this literature is a claim that centralized wage-setting differs from local bargaining because centralized wage-setters take into account the likely policy response of the government to the unions' pay demands. This topic is large and diverse enough to warrant a review of its own.

The literature we did cover does yield a simple, albeit not very exciting, answer to our central question regarding the effect of the degree of centralization on economic performance: It depends. Fortunately, we can say something about what it depends upon. The effect of centralization depends on the industry, the way workers are divided into separate unions and the measure of performance. It matters whether the industry is expanding or declining, or whether firms on the margin of entry or exit are the most productive or the least productive. It also matters whether centralization implies the joining of different types of workers, or the same type of worker in different firms. Finally, it matters whether one is concerned about employment, investment, productivity growth, or equality.

That the effect of centralization depends upon the structure of the industry and the way unions are organized implies a certain scepticism regarding the empirical studies of centralized versus decentralized bargaining that have been done so far. If industry structure matters, then the appropriate test for at least some of the effects of centralization is at the industry, not national, level. If the way workers are divided into unions matters, then one-dimensional indices of centralization are misleading. Countries rank differently along different dimensions of centralization.

Most of the existing indices try to rank countries on a scale that goes from firm-level to industry-level to national-level bargaining. By this measure, Japan is the most decentralized, the United States a little more centralized, the UK a bit above the United States, Germany more centralized than the UK, and Norway and pre-1983 Sweden the most centralized of all. A different measure would rank countries on a scale that goes from every occupation in a separate union to all occupations bargaining jointly. Now the ranking would have the UK as the most decentralized, the

[4] See, among others, Sampson 1983; Calmfors and Horn 1986; Gylfason and Lindbeck 1986; and the papers collected in *Scandinavian Journal of Economics*, 87 (1985). Rødseth 1991 presents a recent appraisal of this literature.

United States a little more centralized, Sweden and Norway in the middle, and Germany and Japan as the most centralized of the six. Should the Japanese case be used as an illustration of the advantages of decentralization or centralization? Clearly we should stop talking about centralization in the singular and analyse the different dimensions of centralization separately.

Finally, whether different dimensions of centralization are beneficial for economic performance depends on the aspect of performance under consideration. Both local bargaining or two-tiered bargaining in which local bargaining adds an increment to a centrally negotiated wage provide incentives for workers to work hard on the job. Bargaining systems without local bargaining do not. Thus concern with effort argues in favour of local bargaining or mixed systems in which central and local bargaining are combined.

With regard to capital formation, however, the results are exactly the reverse. Local bargaining discourages investment as workers at the local level obtain a share of the productivity increase that investment creates. Unfortunately, not all good things go together. Local wage-bargaining as a form of profit-sharing induces the efficient use of inputs whose costs are paid continually, like effort on the job. Local wage-bargaining as a way of sharing current profits but not prior costs, induces too little use of inputs whose costs are sunk, like new plant and equipment. The best bargaining system for capital formation may be the worst for workers' effort. If one is ultimately concerned with, say, per capita GDP, then one's evaluation of different bargaining systems depends on one's assessment of the relative importance of investment in new plant and equipment versus inducing greater effort on the job in the process of economic growth.

The most prominent concern in the literature is with unemployment and it is regarding unemployment that the theoretical results are most diverse. If unions set the wage, if there is one union for each final product, and if prices are sensitive to the quantity of domestic output, then the relationship between centralization and unemployment is hump-shaped with either highly centralized or fully decentralized systems superior to intermediate levels of centralization. If unions set the wage, if there are multiple unions engaged in producing each final product, and if final product prices are fixed in world markets, there is a monotonic relationship with unemployment lower the higher the level of centralization. If unions set the wage, if there is one union per product, if prices are fixed, and if unions care about relative wages or the aggregate rate of unemployment, then highly centralized bargaining entails the lowest unemployment. Highly centralized bargaining also produces the lowest unemployment if wage-setting is driven by union protection of insiders, or by efficiency wage considerations. In contrast, decentralized bargaining produces the lowest unemployment when the wage is set by a process of bargaining between workers who want the highest wage possible and employers provided employers determine employment unilaterally. Thus, in regard to unemployment, there are arguments in favour of both firm-level bargaining and national-level bargaining.

One way to summarize these conflicting results regarding centralization and employment is to distinguish between those arguments that rely on union weakness and those that rely on union co-operation. The advantage of local bargaining in terms of employment depends on the ability of employers to lower the wage by increasing employment. If unions are powerful enough at the local level to prevent employers from increasing the number of insiders, local bargaining would result in less, not more, employment than centralized bargaining. Similarly, the advantage of decentralized bargaining when prices are affected by wage costs disappear if unions are strong enough to organize on an industry basis and obtain the monopoly rents created as an industry moves up its demand curve.

On the other side, the advantages of centralized bargaining depend on real co-operation among the different unions. Centralized bargaining lowers unemployment if unions can agree on how to spread the wage increase among themselves. In the absence of internal agreement among the unions, bargaining that appears to be centralized can become a form of multilateral bargaining that is not centralized at all.

In the presence of strong, cohesive unions, a mixed system of centralized bargaining over the base wage and subsequent firm-level bargaining under a peace clause may be the best compromise between divergent concerns. On the one hand, workers' effort on the job appears to be increasingly important in light of current trends toward more flexible specialization and work organization described by Michael Piore and Charles Sabel (1984). Thus the cost of industry-level or even national-level bargaining without subsequent bargaining at the level of the enterprise may be high. On the other hand, the large difference between labour costs at the level of the firm and labour costs at the level of the national economy points to the existence of a sizeable externality in wage-setting where a substantial part of the costs of wage increases are borne by workers in other unions and other firms.

Separate bargaining by different groups of workers within the firm reduces economic performance on all dimensions. But co-operation among workers divided into blue-collar, white-collar, and professional confederations has proven to be difficult to achieve in the Nordic countries. Indeed, the greatest weakness of national-level bargaining is the difficulty of attaining a consensus among the unions who compete with each other as well as with employers.

The alternative approach, apparently favoured by Swedish employers, is to rely instead on union weakness. The risk with this strategy is that unions would remain strong enough to block the advantages of local bargaining. The danger is that the Nordic countries might lose the advantages of centralized bargaining without obtaining the advantages of decentralized bargaining.

15

Comment

ASSAR LINDBECK

The essay by Moene, Wallerstein, and Hoel is an excellent summary and elaboration of the literature on centralized versus decentralized bargaining. There is no doubt that the authors advance the literature in this field in important respects. My only worry is whether the focus in the literature is reasonable. Let me elaborate.

The main argument for decentralized economic decision-making in national economies is that *the information about time and place* within firms about production processes and markets cannot be centralized without losing the bulk of that information. We would also expect that such a centralization process would *distort* the information. This is, of course, the main argument for decentralized decision-making—i.e. to allow decisions to be taken where the information is available—a point that was first emphasized some fifty years ago, in particular by Friedrich von Hayek.

As the authors observe, recent literature on the advantages of centralized wage-setting neglects such considerations. Instead, the argument has been that centralization is conducive to the internalization of various types of externalities. On the basis of this argument, it is asserted that centralized bargaining is more responsible for society at large than decentralized bargaining, and that this responsible behaviour is an important explanation for the low rate of unemployment in countries with centralized bargaining.

One obvious difficulty with this argument is that some countries that are characterized by centralized bargaining have, in fact, relied heavily on temporary public-sector relief work and permanent increases in public-sector employment to keep down the unemployment rate. They have also used various subsidies to avoid rising unemployment. It is, therefore, natural to ask why countries with assertedly centralized bargaining have not been able to 'deliver' more modest *nominal* wage increases.

Indeed, to the extent that the 'quasi-centralized bargaining' in the Scandinavian countries in the mid-1970s, and the first half of the 1980s, helped reduce product wages, this was *not* usually brought about by nominal wage moderation. The mechanism was rather that unions did not immediately ask for full compensation for price increases in connection with devaluations. This, then, is a much more modest assertion of the 'favourable' effects of the (semi-)centralized bargaining system in the Scandinavian countries than the notion that centralized bargaining in these countries has contributed to limiting the rate of nominal wage increase. After all, devaluations have traditionally been regarded as a method of reducing the overall real wage level in systems with strong unions that engage in *decentralized bargaining*, with rivalry between unions which act independently.

Thus, my claim is that the low levels of unemployment in Sweden, Norway, and Finland in the 1970s and 1980s was the consequence of macroeconomic policies—public-sector employment increases, subsidies, and devaluations—rather than of centralized bargaining. Moreover, these macro policies are not sustainable in the long run.

Another worry with the hypothesis that centralized wage-setting is conducive to wage moderation and low unemployment is that the degree of centralization in wage-bargaining is hardly less in Denmark, with about 10 per cent unemployment in the 1980s, than in the other Nordic countries. And the particularly centralized system of wage-bargaining in Finland did not prevent a 5 per cent rise in unemployment at the end of the 1980s, and to 12 per cent in the early 1990s. Thus, all in all, it seems rather dubious to refer to the Scandinavian countries in support of the view that centralized bargaining is a potentially useful path to wage moderation.

How, then, do these comments on the situation in Scandinavia relate to the argument in the theoretical literature that centralized bargaining takes care of various negative externalities of high nominal wage increases? My answer is that we should be sceptical about these claims. First of all, as pointed out by Calmfors and Driffill (1988), the market power of unions increases through centralized unions, which by itself would be expected to boost wages in a bargaining framework. Second, the argument about the advantages of centralization largely neglects the heterogeneity of both labour and jobs. The argument for centralized bargaining, therefore, often misses the consequences for *relative* wages. Centralization of wage formation means that decisions are made without the decision-makers having access to specific information that is available only at the level of individual firms and/or groups of workers. This may be a reason why centralized bargaining in the Scandinavian countries has resulted in a drastic compression of skill differentials of wages, and, in the judgement of many observers, including myself, has contributed to severe allocation and productivity problems in these countries, in particular perhaps in Sweden. Clearly, there is a risk that such a system of wage formation results in mismatches in the labour market, and hence contributes to unemployment.

The argument that centralized bargaining takes care of various externalities, and hence increases welfare in society, is quite similar to the argument that centralized price-fixing by the government, and indeed central planning in general, is able to take care of externalities and thus be conducive to welfare. The basic weakness of this argument, in both cases, is that *important information and incentive deficiencies of centralized decision-making are left out of the analysis*. This is, in my judgement, the basic weakness of the entire literature on centralized versus decentralized wage-setting—indeed the same weakness as in the Lange–Lerner advocacy of market socialism with centrally determined product prices, as well as in various arguments for government price control à la Galbraith.

I would also suggest that the notion that a country can 'choose' between centralized and decentralized bargaining may be a misspecification of available alternatives. Assuming that wage drift cannot be eliminated, the issue rather seems to be *at how many levels* bargaining should be conducted. If all bargaining is decentralized to the level of individual firms, bargaining will (by definition) be conducted at just *one* level. If bargaining is highly centralized, we would expect it to occur at two or three levels—the central level, the firm level, and possibly also at an intermediate level (of industrial branches).

I would hypothesize that total nominal wage increase is likely, *ceteris paribus*, to be larger if wage formation takes place on two (or even three) levels than if it occurs on just one level. In this sense, centralized bargaining may very well be counter-productive to ambitions to keep down the rate of nominal wage increase, as well as the level of the real product wage.

Moreover, it is not self-evident that centralized bargaining, when *imposed* on a country where bargaining used to be decentralized, would result in the same wage behaviour as in a country that has 'spontaneously' developed centralized wage-bargaining. In the latter type of country, both centralized bargaining and 'responsible' wage increases (to the extent that they actually are responsible) may be joint results of social consensus, rather than the responsible wage behaviour being *caused* by centralized bargaining.

There is, in my view, another even more important weakness of centralized bargaining. In such a system, the central union leadership can exert political blackmail *vis-à-vis* the government. The unions can ask for political favours for themselves as a requirement for their willingness to accept wage moderation. Examples are privileges to strike and picket against a 'third agent', rights to enforce unionization in non-unionized firms, various tax-financed subsidies to unions (such as the tax deductibility of union member fees), or even a tax-financed takeover of the ownership of shares in private firms by the unions (an idea pursued by the Swedish Confederation of Trade Unions in the 1970s and 1980s).

These are some of the reasons why I feel extremely sceptical, or even hostile, to centralized wage-setting. It is probably true, however, that *general discussions* between unions and employers' associations, at levels above individual firms, about the 'reasonable' size of aggregate nominal wage changes, are conducive to the national economy. However, as illustrated by e.g. Japan and Switzerland, and to some extent also West Germany, such discussions do not require centralized wage-bargaining.

16

Comment

LARS CALMFORS

The relation between the extent of centralization of wage-bargaining and economic performance has received much attention recently in the fields of both economics and political science. So a survey of the present state of knowledge, which the contribution by Moene, Wallerstein, and Hoel provides is particularly useful. However, the essay is more than just a survey, since it also addresses a number of highly relevant but hitherto neglected topics. All this is done with the help of a simple and elegant model framework.

An important message is that the results depend upon which goal variable is in focus and against which dimension centralization is measured. I agree with their conclusion that decentralization across industries and firms is likely to have quite different effects from decentralization across professions. This insight is often lost in the public debate. A third dimension—which is not addressed by the authors—is regional decentralization.

The essay consists of two main parts: Chapter 11, which deals with union wage demands, and Chapter 12, which provides an explicit analysis of the wage-bargaining process. I shall comment on both.

The Basic Union Model

Chapter 11 surveys earlier work on how the internalization of the effects of wage increases and the market power of unions influence wage demands at different degrees of centralization. For instance, the authors show that in a simple monopoly union

I am grateful to Gunnar Jonsson for comments and to Helena Matheou for typing and language improvements.

model of a closed economy with endogenous prices and perfect competition in product markets, wage-setting at firm level and complete centralization will lead to the same wage outcome. Industry-level wage-setting will always produce a higher wage (and lower employment) than both these extremes. This is an example of the Calmfors–Driffill (1988) hump-shaped relation between the wage and the extent of centralization. The results can, however, be presented in a more general way than in the essay.

Suppose that the union attached to a specific firm has the utilitarian utility function

$$u = Lv(w_c) + (M - L)v_0, \qquad (1)$$

where u = union utility, L = employment, v = the utility of an employed worker, M = union membership, and v_0 = the utility of an unemployed worker. $w_c = W/Q$ is the real consumption wage, where W = the nominal wage and Q = the consumer price index. The union chooses the real consumption wage w_c that maximizes (1) subject to the labour-demand constraint

$$L = L(w_p), \qquad (2)$$

where $w_p = W/P$ = the real product wage and P = the output price. It follows from the definitions of w_p and w_c, that $w_p = w_c/p_r$, where $p_r = P/Q$ is the relative price between the output of the firm and the consumption basket. The optimization gives rise to the first-order condition

$$\phi(w_c, v_0, \beta, \varepsilon) = w_c v' - \beta(1 - \varepsilon)(v - v_0) = 0, \qquad (3)$$

where $\beta = -(\partial L/\partial w_p) \cdot (w_p/L)$ is the labour-demand elasticity and $\varepsilon = (\partial p_r/\partial w_c) \cdot (w_c/p_r)$ is the elasticity of the relative price of the firm with respect to the real consumption wage of the firm. The elasticity should be interpreted as a *total* elasticity, since it also incorporates the price effects of wage increases in other firms that may be part of the same bargain and therefore occur simultaneously. To simplify, I assume that β is a constant, i.e. that the production function is Cobb-Douglas, and that ε is independent of w_c and p_r. Since the second-order condition for a utility maximum is that $\phi_1 < 0$, and $\phi_4 = \beta(v - v_0) > 0$, it follows that $dw_c/d\varepsilon = -\phi_4/\phi_1 > 0$, i.e. that the larger the price elasticity, the higher the wage. The intuition is that the more the relative price increases when the real consumption wage is raised, the smaller the increase of the real product wage and hence the smaller the employment loss. Therefore the incentive to hold back the wage is reduced.

Consider a symmetrical closed economy consisting of a number of industries, where each industry is made up of a large number of perfectly competitive firms! Obviously firm-level wage-setting and complete centralization then give the same wage, since $\varepsilon = 0$ in both cases. In the former case the reason is that the product price as well as the consumer price index are taken as exogenous, in the latter that the relative price cannot change if the real consumption wage changes equally in all firms.

Equation (3) also explains what happens in between these extremes. Suppose we start with firm-level wage-setting and then let unions aggregate within industries! The larger the fraction of an industry, the unions of which co-operate and raise wages together, the larger the effect on the relative price, i.e. the larger is ε. Hence the wage increases and reaches a maximum when all unions in an industry act in collusion. But when different industry unions raise wages together, the relative-price increase becomes smaller (if the goods produced by the various industries are equally substitutable for each other). Hence we get a continuous hump-shaped relation as in Fig. 16.1. I shall add a few comments to this.

The Bargaining Case

There is no need to restrict the analysis to the monopoly-union case as is done in the essay. Indeed, it is straightforward to extend the model to the bargaining case. In fact, similar forces will work via the employer side and preserve the hump shape in the general case as well. Consider, for example, a Nash bargaining solution, where we assume that the wage is set so as to maximize

$$B = (u - \tilde{u})^{\alpha}(\pi - \tilde{\pi})^{1-\alpha}, \tag{4}$$

where $\pi = p_r Y - w_c L$ is the real profit of the firm, $Y = F(L)$ is the output, and $\tilde{u}$ and $\tilde{\pi}$ are the fall-back positions (the utility levels in case of disagreement) of the union and the firm respectively. If we employ the union utility function (1) and assume that $\tilde{u} = Mv_0$ and $\tilde{\pi} = 0$, the Nash bargaining product becomes

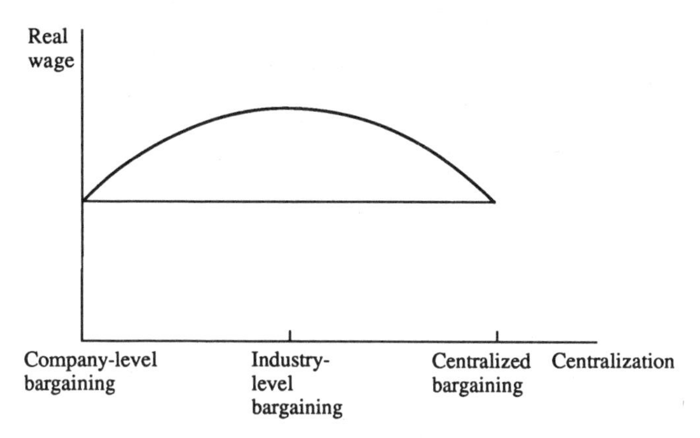

FIG. 16.1 Centralization and Real Wages with Price Externalities

$$B = [L(v(w_c) - v_0)]^{\alpha}\pi^{1-\alpha}. \tag{4'}$$

Maximization subject to (2) gives the first-order condition

$$\gamma(w_c, v_0, \beta, \varepsilon) = \alpha\pi[v'w_c - \beta(1-\varepsilon)(v - v_0)] + (1-\alpha)L(v - v_0)w_c\left[\frac{\beta\varepsilon}{\beta - 1} - 1\right] = 0, \tag{5}$$

where the second-order condition for a maximum ensures that $\gamma_1 < 0$. It again follows that a larger price elasticity increases the wage, since $\gamma_4 = \alpha\pi\beta(v - v_0) + (1-\alpha)L(v - v_0)\beta w_c/(\beta - 1) > 0$ and hence $dw_c/d\varepsilon = -\gamma_4/\gamma_1 > 0$. The first part of γ_4 captures the same effect as before, the second that the fall in profits induced by a wage increase is smaller the more the relative output price can be raised.

The Open-Economy Case

The above analysis applies to a closed economy. Moene *et al.* do, however, also treat the case of an open economy, where all (relative) prices are determined in the world market. Then $\varepsilon = 0$ at all levels of centralization. Thus it follows that the wage will be the same independently of the level of centralization. The hump-shaped relation between the wage and the extent of centralization degenerates into a horizontal line (I in Fig. 16.2).

However, a more relevant open-economy case is when the domestic economy produces goods that are imperfect substitutes

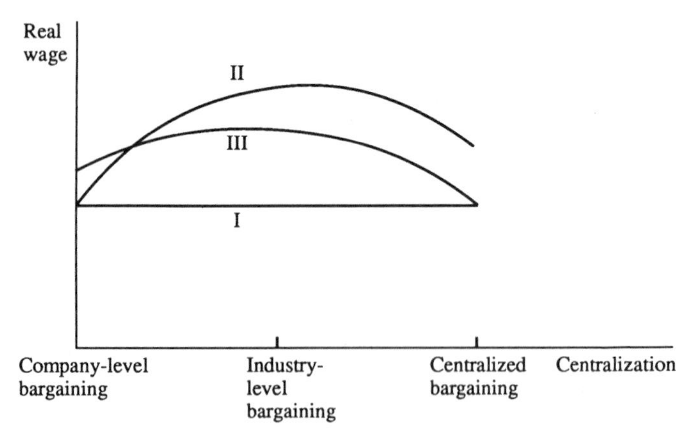

FIG. 16.2 Possible Relations between Centralization and Real Wages

for those produced abroad, and where increases in the domestic real consumption wage raises the price of domestic goods relative to foreign goods. We can think of the world economy as consisting of a number of production sectors, some of which are located in the domestic economy and some in the rest of the world. Increasing centralization of wage-setting in the domestic economy will then as before give a hump-shaped relation. But there will be one major difference. Complete decentralization to the firm level and complete centralization in the domestic economy no longer give the same outcome. $\varepsilon = 0$ now holds only in the first case; with complete centralization $\varepsilon > 0$, since equal wage increases throughout the domestic economy will now raise domestic prices relative to foreign. Hence complete decentralization now produces a lower wage than complete centralization in this set-up (curve II in Fig. 16.2).

Other Considerations

The above analysis has been cast in terms of perfectly competitive firms. Moene *et al.* also consider explicitly monopolisitic competition and show that the increase of the product real wage is counteracted by a relative-price increase under firm-level wage-setting as well in this case. This tends to give a lower wage with complete centralization than with complete decentralization. Other forces working in the same direction are the internalization effects on, for instance, taxes, aggregate employment, 'envy', and effort of workers, which are also discussed in the essay.

Summarizing all aspects, it does indeed seem probable that complete centralization produces lower wages than complete decentralization even in an open economy (curve III in Fig. 16.2). Still, it is an open question how relevant this conclusion really is, since complete centralization may not be a relevant alternative in the real world. The relevant 'choice' may rather be between decentralized systems and semi-centralized ones (with a small number of bargaining units on each side). The semi-centralized systems will fail to realize the full benefits of internalization: it may make a large difference whether, say, *all* tax effects are internalized or only 1/2 or 1/3. Hence the differences in the extent of internalization between the systems that are of practical relevance may not be of so large an importance as is often taken for granted (Calmfors 1993).

The Moene *et al.* Analysis of Bargaining

Chapter 13 of the essay deals explicitly with bargaining, although this is done within a different context than in the earlier parts: the unions are now assumed to be interested in the wage only and the earlier analysed effects via prices, employment, taxes, 'wage jealousy', etc. no longer play any role. Instead, effects working via entry and exit of firms, 'collective effort', investment, and the employment decisions of firms are stressed. Although this part is less homogeneous than the survey in Chapter 12, it does provide a number of novel insights. One straightforward—but hitherto neglected—conclusion is that employment can be higher with decentralized than with centralized bargaining, since it may then pay employers to expand employment beyond the point at which the value of the marginal product of labour equals the wage. The reason is that increased employment will not only raise output, but also (with the assumptions made on bargaining) reduce the wage to the extent that output per worker is reduced.

Wage Drift and Central Contracts

The implicit assumption in most analyses of bargaining structure—including my own—have been that central bargainers can control the final wage outcome. This may, however, be an inappropriate assumption since in practice centralized bargaining is a mixed system, which presupposes bargaining both at the central level and at lower levels. Moene *et al.* make an important contribution by trying explicitly to sort out the relation between wage drift and central contract wages. This is done under the assumption that employment is determined prior to the bargaining.

The Rubinstein (1982) interpretation of the Nash bargaining solution gives the bargained wage in an individual firm as a weighted average of the best and worst outcomes that the union might achieve. With the assumption that union utility equals the real wage, one obtains

$$W = \alpha \frac{R - \tilde{\Pi}}{L} + (1 - \alpha)\tilde{W}, \tag{6}$$

where W = the money wage, R = the revenue of the firm, $\tilde{\Pi}$ = the money profit of the firm during a conflict, L = employment, and $\tilde{W}$ = worker income during a conflict.[1] Moene *et al.* consider two cases. In the first, strikes and lock-outs are allowed at the local level. There is no outside conflict support. Then $\tilde{\Pi} = \tilde{W} = 0$ and hence

$$W = \alpha \frac{R}{L} \tag{7a}$$

and

$$d = \alpha \frac{R}{L} - q, \tag{7b}$$

where d = wage drift, and q = the central contract money wage. Thus, according to (7a) the final money wage is a fraction of the value of average labour productivity. Moreover, the local negotiations determine the final wage outcome. According to (7b) drift will offset any deviation of the contract wage from the 'desired' outcome at the local level.

In the other case considered, a formal peace clause applies to

[1] The formula holds under the assumption of perfect competition, in which case both the product price of the firm and the consumer price index can be taken as exogenously given.

local bargaining. However, in practice, this means that employers are obliged to pay the contract wage, whereas workers can inflict damage on the employer by 'go-slow' actions, etc. (it is assumed that they can reduce output by a fraction θ). Hence $\tilde{\Pi} = \theta R - qL$ and $\tilde{W} = q$. This gives

$$W = \alpha(1-\theta)\frac{R}{L} + q \tag{8a}$$

and

$$d = \alpha(1-\theta)\frac{R}{L}. \tag{8b}$$

According to (8b) there will *always* be wage drift. It will be a fraction of the value of average labour productivity that becomes larger the more workers are able to reduce output (the smaller is θ). However, since drift is independent of the central wage, central bargainers will always be able to achieve their desired final wage outcome by taking the subsequent drift into account (equation 8b). This particular case seems best to describe the conditions in the Nordic countries. The analysis motivates a few comments.

Variations of the Theme

(i) The result that there will always be wage drift in the second case follows from the asymmetry that workers can inflict damage on the firm, whereas the firm is not allowed to reduce pay below the contract wage. Suppose that the firm would have to pay only a fraction δ of the central contract wage in case local negotiations break down. Then $\tilde{W} = \delta q$ in equation (6) gives

$$W = \alpha(1-\theta)\frac{R}{L} + \delta q \tag{9a}$$

and

$$d = \alpha(1-\theta)\frac{R}{L} - (1-\delta)q. \tag{9b}$$

Obviously δ can be chosen so that all drift is eliminated. Suppose that the central contract wage is set so that $q = \alpha R/L$.[2] Then $\delta = \theta$ is a necessary and sufficient condition for zero wage drift. The condition implies that there must be the same reduction of workers' pay and of output (and thus also of the profit).

(ii) The conclusion that central bargainers can determine the final wage outcome does not presuppose that workers receive the full contract wage in the case of local disagreement. According to (9a), it is enough that the wage paid out during a local conflict depends upon the contract wage ($0 < \delta \leq 1$). Although the central contract wage is always counteracted by a reduction of drift, the offset will not be complete as long as $\delta > 0$.

(iii) A basic assumption in the Moene *et al.* analysis is that central negotiations always precede local ones. Perhaps one could just as well claim the opposite. Also, suppose that we assume limited conflicts at the central level of the same type as is assumed at the local level: workers continue to receive the pay they achieved after the last round of local negotiations but can inflict some damage on the employers. This may not be too unrealistic, since disagreements at the central level often give rise to only limited conflicts. This change of assumptions would turn the conclusions from equations (8a) and (8b) upside down: there will always be a central wage increase, but local bargainers will decide the final outcome.

(iv) However, the most crucial assumption is that there are no constraints on the choice of the central wage. This is the reason why the number of bargaining levels does not matter for the final wage outcome. A reasonable real-world restriction would seem to be that the *money* wage cannot be reduced at the central level (under 'normal' conditions anyway), as has been claimed by Holden (1991) and Calmfors (forthcoming). Because of a desire on the part of central bargainers to influence relative wages e.g., to guarantee money wage increases for those who do not get drift and to show the members that the central union does something for them, there is even, perhaps, a lower bound on the rate of increase of the central money wage. If so, the number of bargaining levels will matter.

[2] This is the final wage outcome desired also by central wage setters if $\tilde{\Pi} = \tilde{W} = 0$ in the case of a conflict at the central level (cf. equation (7a)).

A possible conclusion in a more general model is that mixed bargaining may, indeed, produce real wage restraint because of internalization effects, but only under conditions of inflation. The higher the rate of inflation, the more room there is for central bargainers to adjust central wage increases to the subsequent drift in order to achieve their desired *real* wage outcome. This view would seem consistent with the combination of high inflation and real wage restraint in the Nordic countries in the late 1970s and early 1980s (Calmfors & Nymoen 1990). This favourable outcome would, however, break down with a transition to low inflation, as we seem to be observing. Then central bargainers may fail to reduce central money wage increases to such a low level that they can offset subsequent wage drift. The consequence is that they may be unable to achieve their real wage targets, and that real wage increases may therefore exceed the growth rate consistent with stable employment. Hence, in a low-inflation economy a reduction of the number of bargaining levels may be crucial if continued real-wage restraint is to be achieved. An intriguing question is whether it may be possible to achieve the internalization benefits of centralized bargaining with a semi-decentralized system, where all bargaining takes place at the firm level but where conflict rights rest with the central organizations.

Conclusions

A general conclusion of the essay is that the net effects of centralization versus decentralization appear rather uncertain. Nor is the extent of decentralization a variable that policy-makers can control. This raises the question whether economists have devoted too much interest to this issue from a policy point of view. Perhaps, measures that affect the relative bargaining strength of the parties of the labour market (the ability to inflict damage on the other party and the ability to endure conflicts) are a more certain way of influencing wage outcomes. Such measures have predictable effects under any degree of centralization. I am thinking of such variables as the fines for wild-cat strikes, restrictions on conflicts that may threaten society's vital interests, tax deductibility for union dues, individual contributions to the unemployment insurance, the legal bargaining monopoly of unions for non-members as well for many issues, the legal rights for non-organized labour and firms with non-organized workers, etc. Institutional changes that weaken relative union strength in bargaining are also likely to have profound effects on the political strength of unions. In my view, this may not be a bad thing in a society like the Swedish one, where corporatist influences on policy-making have been strong.

References for Part II

Akerlof, G. A., and Yellen, J. L. (eds.) (1986), *Efficiency Wage Models of the Labour Market*. Cambridge: Cambridge University Press.

Alvarez, M., Garrett, G., and Lange, P. (1991), 'Government Partisanship, Labor Organization and Macroeconomic Performance, 1967–1984', *American Political Science Review*, 85: 539–56.

Andersen, T. M., and Risager, O. (1990), 'Wage Formation in Denmark', in Lars Calmfors (ed.), *Wage Formation and Macroeconomic Policy in the Nordic Countries*. Oxford: Oxford University Press.

Barth, E. (1991), 'Disagreement Points in Wage Bargaining'. Mimeo.: Institute for Social Research, Oslo. 22 Mar. 1991.

Bean, C., Layard, R., and Nickell, S. (1986), 'The Rise in Unemployment: A Multi-Country Study', *Economica*, 53: S1–S22.

Blaas, W. (1982), 'Institutional Analysis of Stagflation', *Journal of Economic Issues*, 16: 955–75.

Blair, D. H., and Crawford, D. L. (1984), 'Labor Union Objectives and Collective Bargaining', *Quarterly Journal of Economics*, 99: 547–66.

Blanchard, O., and Summers, L. (1987), 'Hysteresis in Unemployment', *European Economic Review*, 31: 288–95.

Blinder, A. S. (ed.) (1990), *Paying for Productivity*. Washington, DC: Brookings Institution.

Bowles, S. (1985), 'The Production Process in a Competitive Economy: Walrasian, Neo-Hobbesian and Marxian Models', *American Economic Review*, 75: 16–36.

Bratt, C. (1986), *Labour Relations in 18 Countries*. Stockholm: SAF.

Brunello, G., and Wadhwani, S. (1989), 'The Determinants of Wage Flexibility in Japan: Some Lessons from a Comparison with the UK Using Micro-Data'. Centre for Labour Economics, Working Paper no. 1116. London: London School of Economics.

Bruno, M., and Sachs, J. (1985), *The Economics of Worldwide Stagflation*. Cambridge, Mass.: Harvard University Press.

Cahuc, P. (1987), 'Centralized versus Decentralized Wage Setting: Inflation and Unemployment Implications'. MS. Paris: University of Paris.

Calmfors, L. (1993), 'The Extent of Centralization of Wage Bargaining and Macroeconomic Performance'. MS. Paris: OECD.

—— (forthcoming), 'Lessons from the Macroeconomic Experience of Sweden', *European Journal of Political Economics*.

Calmfors, L., and Driffill, J. (1988), 'Bargaining Structure, Corporatism and Macroeconomic Performance', *Economic Policy*, 6: 13–62.

—— and Horn, H. (1986), 'Employment Policies and Centralized Wage-Setting', *Economica*, 53: 281–302.

—— and Nymoen, R. (1990), 'Real Wage Adjustments and Employment Policies in the Nordic Countries', *Economic Policy*, 11: 379–448.

Calvo, G. A. (1979), 'Quasi-Walrasian Theory of Unemployment', *American Economic Review*, 69: 102–7.

—— and Wellisz, S. (1978), 'Supervision, Loss of Control and the Optimum Size of the Firm', *Journal of Political Economy*, 86: 943–52.

Cameron, D. R. (1984), 'Social Democracy, Corporatism and Labor Quiescence: The Representation of Economic Interest in Advanced Capitalist Society', in J. H. Goldthorpe (ed.), *Order and Conflict in Contemporary Capitalism*. Oxford: Oxford University Press: 143–78.

De Menil, G. (1971), *Bargaining: Monopoly Power Versus Union Power*. Cambridge, Mass.: MIT Press.

Dell'Aringa, C., and Lodivici, M. S. (1990), 'Industrial Relations and Economic Performance', *Review of Economic Conditions in Italy*, 1990: 55–83.

Elster, J. (1989), *The Cement of Society*. Cambridge: Cambridge University Press.

Elvander, N. (1988), 'The Double Challenge to the Trade Unions: Alternative Forms of Remuneration and Internationalization'. MS. Stockholm.

—— (1989), 'Pay System Practices and Labour Flexibility in Sweden'. MS. Stockholm.

Farber, H. S. (1986), 'The Analysis of Union Behavior', in Orley C. Ashenfelter and Richard Layard (eds.), *Handbook of Labor Economics*, ii. Amsterdam: North Holland: 1039–89.

Fernandez, R., and Glazer, J. (1991), *American Economic Review*, 81: 240–52.

Flanagan, R. J. (1990), 'Centralized and Decentralized Pay Determination in Nordic Countries', in Lars Calmfors (ed.), *Wage Formation and Macroeconomic Policy in the Nordic Countries*. Oxford: Oxford University Press.

Freeman, R. B. (1988), 'Labour Market Institutions and Economic Performance', *Economic Policy*, 3: 64–80.

Garrett, G., and Lange, P. (1986), 'Economic Growth in Capitalist Democracies, 1974–1982', *World Politics*, 38: 517–45.

Gottfries, N., and Horn, H. (1986), 'Wage Formation and the Persistence of Unemployment', *Economic Journal*, 97: 877–84.

Grout, P. A. (1983), 'Investment and Wages in the Absence of Legally Binding Labour Contracts: A Nash Bargaining Approach', *Econometrica*, 52: 449–60.

Gylfason, T., and Lindbeck, A. (1984), 'Union Rivalry and Wages: An Oligopolistic Approach', *European Economic Review*, 24: 1–24.

—— —— (1986), 'Endogenous Unions and Governments: A Game-Theoretic Approach', *European Economic Review*, 30: 5–26.

Hall, R. E., and Lilien, D. M. (1979), 'Efficient Wage Bargains under Uncertain Supply and Demand', *American Economic Review*, 69: 868–79.

Haller, H., and Holden, S. (1990), 'A Letter to the Editor on Wage Bargaining', *Journal of Economic Theory*, 52: 232–6.

Headey, B. W. (1970), 'Trade Unions and National Wages Policies', *Journal of Politics*, 32: 407–39.

Heitger, B. (1987), 'Corporatism, Technological Gaps and Growth in OECD Countries', *Weltwirtschaftliches Archiv*, 123: 463–73.

Hersoug, T. (1985), 'The Importance of Being Unimportant: On Trade Unions' Strategic Position'. Memorandum from the Department of Economics no. 12. Oslo: University of Oslo.

—— Kjær, K. N., and Rødseth, A. (1986), 'Wages, Taxes and the Utility-Maximizing Trade Union: A Confrontation with Norwegian Data', *Oxford Economic Papers*, 38: 403–23.

Hibbs, D. A., Jr. (1976), 'Industrial Conflict in Advanced Industrial Societies', *American Political Science Review*, 70: 1033–58.

—— and Locking, H. (1991), 'Wage Compression, Wage Drift, and Wage Inflation in Sweden'. FIEF Working Paper no. 87. Stockholm: Trade Union Institute for Economic Research (FIEF).

Hicks, A. (1988), 'Social Democratic Corporatism and Economic Growth'. *Journal of Politics*, 50: 677–704.

Hicks, J. R. (1963), *The Theory of Wages*. London: Macmillan.

Hoel, M. (1989*a*), 'Efficiency Wages and Local Versus Central Wage Bargaining', *Economics Letters*, 30: 175–9.

—— (1989*b*), 'Employment under Collective Wage Bargaining when there are Hiring and Firing Costs'. Memorandum of the Department of Economics no. 7. Oslo: University of Oslo.

—— (1990), 'Local Versus Central Wage Bargaining with Endogenous Investments', *Scandinavian Journal of Economics*, 92/3: 453–69.

—— (1991), 'Union Wage Policy: The Importance of Labour Mobility and the Degree of Centralization', *Economica*, 58: 139–53.

Holden, S. (1988), 'Local and Central Wage Bargaining', *Scandinavian Journal of Economics*, 90/1: 93–9.

—— (1989), 'Wage Drift and Bargaining: Evidence from Norway', *Economica*, 56: 419–32.

—— (1990), 'Wage Drift in Norway', in Lars Calmfors (ed.), *Wage Formation and Macroeconomic Policy in the Nordic Countries*. Oxford: Oxford University Press.

Holden, S. (1991*a*), 'A Bargaining Theory of Inflation and Unemployment'. MS. Department of Economics, University of Oslo.
—— (1991*b*), 'Economic Policy in an Economy with Local and Central Wage Bargaining'. Memorandum from the Department of Economics no. 8. Oslo: University of Oslo.
—— and Raaum, O. (1989), 'Wage Moderation and Union Structure'. MS. Oslo: SAF Center for Applied Research at the Department of Economics, University of Oslo.
Holmlund, B. (1986), 'Centralized Wage Setting, Wage Drift, and Stabilization Policies under Trade Unionism', *Oxford Economic Papers*, 38: 243–58.
—— and Skedinger, P. (1990), 'Wage Bargaining and Wage Drift: Evidence from the Swedish Wood Industry', in Lars Calmfors (ed.), *Wage Formation and Macroeconomic Policy in the Nordic Countries*. Oxford: Oxford University Press.
Horn, H., and Wolinsky, A. (1988), 'Worker Substitutability and Patterns of Unionization', *Economic Journal*, 98: 484–97.
Ingham, G. (1974), *Strikes and Industrial Conflict: Britain and Scandinavia*. London: Macmillan.
Jackman, R. (1990), 'Wage Formation in the Nordic Countries Viewed from an International Perspective', in Lars Calmfors (ed.), *Wage Formation and Macroeconomic Policy in the Nordic Countries*. Oxford: Oxford University Press: 289–322.
—— Pissarides, C., and Savouri, S. (1990), 'Labour Market Policies and Unemployment in the OECD', *Economic Policy*, 5: 450–90.
Johansen, L. (1982), 'Cores, Aggressiveness and the Breakdown of Cooperation in Economic Games', *Journal of Economic Behavior and Organization*, 3: 1–37.
Jun, B. H. (1989), 'Non-Cooperative Bargaining and Union Formation', *Review of Economic Studies*, 56: 59–76.
Kalai, E. (1985), 'Solutions to the Bargaining Problem', in L. Hurwicz, D. Schmeidler, and H. Sonnenschein (eds.), *Social Goals and Social Organizations: Essays in Memory of Elisha Pazner*. Cambridge: Cambridge University Press.
Kalecki, M. (1943), 'Political Aspects of Full Employment', *Political Quarterly* (Oct.–Dec.): 322–31.
Kalleberg, A. L., and Colbjørnsen, T. (1990), 'Unions and the Structure of Earnings Inequality: Cross-National Patterns', *Social Science Research*, 19: 348–71.
Katzenstein, P. (1985), *Small States in World Markets: Industrial Policy in Europe*. Ithaca, NY: Cornell University Press.
Korpi, W., and Shalev, M. (1980), 'Strikes, Power and Politics in Western Nations, 1900–1976', *Political Power and Social Theory*, 1: 301–34.

Lange, P., and Garrett, G. (1985), 'The Politics of Growth', *Journal of Politics*, 47: 792–827.
Layard, R., Nickell, S., and Jackman, R. (1991), *Unemployment: Macroeconomic Performance and the Labour Market*. Oxford: Oxford University Press.
Lehmbruch, G. (1977), 'Liberal Corporatism and Party Government', *Comparative Political Studies*, 10: 91–126.
—— (1979), 'Concluding Remarks: Problems for Future Research on Corporatist Intermediation and Policy-Making', in Philippe C. Schmitter and Gerhard Lehmbruch (eds.), *Trends Toward Corporatist Intermediation*. Beverly Hills, Calif.: Sage Publications.
Leontief, W. (1946), 'The Pure Theory of a Guaranteed Annual Wage Contract', *Journal of Political Economy*, 54: 76–9.
Lindbeck, A., and Snower, D. J. (1988), *The Insider–Outsider Theory of Employment and Unemployment*. Cambridge, Mass.: MIT Press.
Lutz, M. A. (1981), 'Stagflation as an Institutional Problem', *Journal of Economic Issues*, 15: 745–68.
McCallum, J. (1983), 'Inflation and Social Consensus in the Seventies', *Economic Journal*, 93: 784–805.
—— (1986), 'Unemployment in OECD Countries in the 1980s', *Economic Journal*, 96: 942–60.
McDonald, I. M., and Solow, R. M. (1981), 'Wage Bargaining and Employment', *American Economic Review*, 71: 896–908.
Marks, G. (1986), 'Neocorporatism and Incomes Policy in Western Europe and North America', *Comparative Politics*, 18: 253–77.
Moene, K. O. (1986), 'Local Wage Bargaining and "Recession Resistance"'. Memorandum from the Department of Economics no. 9. Oslo: University of Oslo.
—— (1988), 'Union Threats and Wage Determination', *Economic Journal*, 98: 471–543.
—— (1990), 'Union Militancy and Plant Design', in M. Aoki, B. Gustafsson, and O. Williamson (eds.), *The Firm as a Nexus of Treaties*. Beverly Hills, Calif.: Sage Publications.
—— and Wallerstein, M. (1991*a*), 'Collective Bargaining versus Workers' Ownership'. Institute of Industrial Relations, Working Paper no. 198. Los Angeles, Calif.: UCLA. Forthcoming, *Journal of Comparative Economics*.
—— —— (1991*b*), 'Centralized and Local Wage Bargaining with Entry and Exit'. MS. Los Angeles, Calif.: Department of Political Science, UCLA.
—— —— (1991*c*), 'Solidaristic Bargaining'. MS. Los Angeles, Calif.: Department of Political Science, UCLA.
—— —— (1992), 'Unions versus Cooperatives', in S. Bowles, H. Gintis, and B. Gustafsson (eds.), *The Microfoundations of Political Economy:*

Problems of Participation, Democracy and Efficiency. Cambridge: Cambridge University Press.

Mortensen, D. T. (1986), 'Job Search and Labor Market Analysis', in O. C. Ashenfelter and R. Layard (eds.), *Handbook of Labor Economics*, ii. Amsterdam: North Holland: 849–919.

Myrdal, H. G. (1991), 'The Hard Way for a Centralized to a Decentralized Industrial Relations System: The Case of Sweden and SAF', in O. Jakobi and D. Sadowski (eds.), *Employers' Associations in Europe: Policy and Organization*. Baden-Baden: Nomos Verlag.

Nash, J. F., Jr. (1950), 'The Bargaining Problem', *Econometrica*, 18: 155–62.

Newell, A., and Symons, J. S. V. (1987), 'Corporatism, Laissez-Faire and the Rise in Unemployment', *European Economic Review*, 31: 567–614.

Nymoen, R. (1991), 'A Small Linear Model of Wage and Price Inflation in the Norwegian Economy', *Journal of Applied Econometrics*, 6: 255–69.

Organization for Economic Cooperation and Development (1977), *Towards Full Employment and Price Stability*. Paris: OECD.

Oswald, A. J. (1979), 'Wage Determination in an Economy with Many Trade Unions', *Oxford Economic Papers*, 31: 369–85.

—— (1982), 'The Microeconomic Theory of the Trade Union'. *Economic Journal*, 92: 576–95.

—— (1985), 'The Economic Theory of the Trade Unions: An Introductory Survey', *Scandinavian Journal of Economics*, 87: 160–93.

—— (1987), 'Efficient Contracts are on the Labour Demand Curve: Theory and Facts', Centre for Labour Economics, Discussion Paper no. 284. London: London School of Economics.

Paldam, M., and Pedersen, P. J. (1984), 'The Large Pattern of Industrial Conflict—A Comparative Study of 18 Countries 1919–1979', *International Journal of Social Economics*, 11/5: 3–28.

Paloheimo, H. (1990), 'Micro Foundations and Macro Practice of Centralized Industrial Relations', *European Journal of Political Research*, 18: 389–406.

Piore, M., and Sabel, C. (1984), *The Second Divide: Possibilities for Prosperity*. New York: Basic Books.

Pohjola, M. (1984), 'Union Rivalry and Economic Growth: A Differential Game Approach', *Scandinavian Journal of Economics*, 86: 365–70.

—— 'Corporatism and Wage Bargaining'. Discussion Paper no. 85. Helsinki: Labour Institute for Economic Research.

Raaum, O. (1990), 'Centralized Versus Local Wage-Setting with Insider Unions', MS. Oslo: SAF Centre for Applied Research, University of Oslo.

Rosen, S. (1970), 'Unionism and the Occupational Wage Structure in the United States', *International Economic Review*, 11: 269–86.

Ross, A. M., and Hartman, P. T. (1960), *Changing Patterns of Industrial Conflict*. Berkeley, Calif.: University of California Press.

Roth, A. E. (1979), *Axiomatic Models of Bargaining*. Berlin: Springer-Verlag.

Rowthorn, R. (1989*a*), 'Social Corporatism, Wage Dispersion and Labour Market Performance', MS. Cambridge: Economics Faculty, University of Cambridge.

—— (1989*b*), 'Wage Dispersion and Employment: Theories and Evidence', MS. Cambridge: Economics Faculty, University of Cambridge.

Rødseth, A. (1990), 'Efficiency Wages and Local versus Central Bargaining'. MS. Oslo: University of Oslo.

—— (1991), 'Are Employment Policies Counterproductive when Wage Setting is Centralized?' MS. Oslo: University of Oslo.

—— and Holden, S. (1990), 'Wage Formation in Norway', in Lars Calmfors (ed.), *Wage Formation and Macroeconomic Policy in the Nordic Countries*. Oxford: Oxford University Press.

Rubinstein, A. (1982), 'Perfect Equilibrium in a Bargaining Model', *Econometrica*, 50.

Sampson, A. (1983), 'Employment Policy in a Model with a Rational Trade Union', *Economic Journal*, 93: 297–311.

Schmitter, P. C. (1974), 'Still the Century of Corporatism?' *Review of Politics*, 36: 7–52.

—— (1977), 'Modes of Interest Intermediation and Models of Societal Change in Western Europe', *Comparative Political Studies*, 10: 7–38.

Shaked, A., and Sutton, J. (1984), 'Involuntary Unemployment as a Perfect Equilibrium in a Bargaining Model', *Econometrica*, 52: 1351–64.

Shapiro, C., and Stiglitz, J. (1984), 'Equilibrium Unemployment as a Worker Discipline Device', *American Economic Review*, 74: 433–44.

Soskice, D. (1990), 'Wage Determination: The Changing Role of Institutions in Advanced Industrial Countries', *Oxford Review of Economic Policy*, 6: 36–61.

Ståhl, I. (1972), *Bargaining Theory*. Stockholm: Economics Research Institute.

Strand, J. (1989), 'Oligopoly with Monopoly Unions'. Memorandum from the Department of Economics no. 10. Oslo: University of Oslo.

Summers, L. L. (1988), 'Relative Wages, Efficiency Wages and Keynesian Unemployment', *American Economic Review: Papers and Proceedings*, 78: 383–9.

Sutton, J. (1986), 'Non-Cooperative Bargaining Theory: An Introduction', *Review of Economic Studies*, 53: 709–24.

Tarantelli, E. (1986), 'The Regulation of Inflation and Unemployment', *Industrial Relations*, 25: 1–15.

Therborn, G. (1987), 'Does Corporatism Really Matter? The Economic Crisis and Issues of Political Theory', *Journal of Public Policy*, 7: 259–84.

Tyrväinen, T. (1989), 'Why Break Down Centralized Wage Bargaining?' Discussion Paper 22/89. Helsinki: Bank of Finland.

Uddén-Jondal, E. (1989), 'Wage Formation in a Unionized Economy with Two Groups of Workers'. Seminar Paper no. 435. Stockholm: Institute for International Economic Studies.

—— (1990), 'Wage Formation and the Impact of Envy'. MS. Stockholm: Trade Union Institute for Economic Research (FIEF).

Wallerstein, M. (1987), 'Unemployment, Collective Bargaining and the Demand for Protection', *American Journal of Political Science*, 31: 729–52.

—— (1990), 'Centralized Bargaining and Wage Restraint', *American Journal of Political Science*, 34: 982–1004.

Weitzman, M. (1983), 'Some Macroeconomic Implications of Alternative Compensation Systems', *Economic Journal*, 93: 763–83.

—— (1984), *The Share Economy: Conquering Stagflation*. Cambridge, Mass.: Harvard University Press.

Index

Index compiled by Frank Pert